THE PIVOTAL GENERATION

The Pivotal Generation

Why We Have a Moral Responsibility to Slow Climate Change Right Now

Henry Shue

PRINCETON UNIVERSITY PRESS

PRINCETON AND OXFORD

Published by Princeton University Press
41 William Street, Princeton, New Jersey 08540
6 Oxford Street, Woodstock, Oxfordshire OX20 1TR

press.princeton.edu

All Rights Reserved

Library of Congress Cataloging-in-Publication Data

Names: Shue, Henry, author.
Title: The pivotal generation : why we have a moral responsibility
    to slow climate change right now / Henry Shue.
Description: Princeton : Princeton University Press, [2021] |
    Includes bibliographical references and index.
Identifiers: LCCN 2021013539 (print) | LCCN 2021013540 (ebook) |
    ISBN 9780691226248 (hardback ; acid-free paper) | ISBN 9780691226255 (ebook)
Subjects: LCSH: Climate change mitigation. | Environmental policy—United
    States. | Environmental ethics. | Climatic changes—Forecasting.
Classification: LCC TD171.75 .S58 2021 (print) | LCC TD171.75 (ebook) |
    DDC 363.738/7460973—dc23
LC record available at https://lccn.loc.gov/2021013539
LC ebook record available at https://lccn.loc.gov/2021013540

Editorial: Rob Tempio and Matt Rohal
Production Editorial: Natalie Baan
Jacket Design: Pamela L. Schnitter
Production: Erin Suydam
Publicity: Alyssa Sanford and Amy Stewart

Jacket art: Shutterstock

This book has been composed in Adobe Text and Gotham

Printed on acid-free paper. ∞

Printed in the United States of America

10 9 8 7 6 5 4 3 2 1

For the
Augusta County Alliance
(now, Alliance for the Shenandoah Valley)
and all the other grassroots Davids
who slew the dangerous Goliath
that would have been the
Atlantic Coast Pipeline

Different motives, one goal
Coalition politics

CONTENTS

This book is intended primarily for US citizens who would like to think systematically about our responsibility, both simply as humans and also specifically as Americans, to confront climate change. Many of the arguments apply equally well to people in other affluent nations, but in order to stay concrete and not lapse into vague platitudes, I have mostly stuck to this one case of my own country. Because systematic thought about responsibility requires venturing into political and ethical philosophy, I hope the book will interest philosophers and students of philosophy without being inaccessible to nonphilosophers.

I intend this book to be practical philosophy. So, on the one hand, I have avoided technical philosophical issues with no implications for how to act. On the other hand, I have waded into crucial scientific and political judgments to the best of my limited ability. Philosophy can be practical only by being empirically embedded, because right action in the world can never be specified purely by conceptual arguments.

Around 1990 at Cornell University, Duane Chapman insisted that I ought to look into the ethical issues about climate change; I am still wrestling with them. Andy Hurrell issued the invitation that led me to write "The Unavoidability of Justice" (1992), and Fred Aman invited what became "Subsistence Emissions and Luxury Emissions" (1993). Over the years, I have learned and received encouragement from Steve Gardiner, Dale Jamieson, Darrel Moellendorf, Tim Hayward, Liz Cripps, Anja Karnein, and Simon Caney. Rob Tempio at Princeton University Press suggested this book, and Catriona McKinnon gave the manuscript an invaluably constructive critique. I also learned from another anonymous reader for the Press.

For the most conversations over the most years by far, I am grateful to Vivienne, who has continued not merely to tolerate but strongly to support this work during what we quizzically and smilingly call our "retirement."

Witney, England
January 2021

THE PIVOTAL GENERATION

1

The Pivotal Generation

"YOU ARE HERE"

The crucial role we fill, as moral beings, is as members of a cross-generational community, a community of beings who look before and after, who interpret the past in the light of the present, who see the future as growing out of the past, who see themselves as members of enduring families, nations, cultures, traditions.[1]

We continue to live on a stage where there is nothing but the present. Past and future alike have dissolved into a perpetual now, leaving us imprisoned in a moment without links backwards or forwards.[2]

Every decade is consequential in its own way, but the twenty-twenties will be consequential in a more or less permanent way. Global CO_2 emissions are now so high—in 2019, they hit a new record of forty-three billion metric tons—that ten more years of the same will be nothing short of cataclysmic.[3]

Illusions of Separation

Climate change is a matter of time. As we ordinarily think of time, now is the critical time for vigorous action to try to impose some limit on climate change. Human action or inaction during the next decade or so is likely to determine how severe climate change finally becomes. It is still—only barely—possible for us to act just in time to prevent the worst in spite of the fact that invaluable time has been thrown away by callous and corrupt political leaders who have largely wasted the last three decades since the Framework Convention on Climate Change was adopted with much fanfare in 1992 and by the executives in the fossil-fuel industry who have deceived and tricked the public and corrupted our politics so that they can continue business-as-usual for as long as possible.[4] That we still have the opportunity to act just in time makes us here and now the most important generation of humans to have lived with regard to the conditions of life on this planet for us and all the other species. We can be the "greatest generation" for the climate struggle or the miserably self-preoccupied and easily manipulated ones who failed to rise to the occasion and whom future generations will recall, if at all, with contempt. "Time is of the essence," as the lawyers phrase it in the contracts. The time is now, and the time is short. So those of us alive now are the pivotal generation in human history for the fate of our planet's livability.

Yet climate change is also a matter of time in a deeper, more philosophically interesting and morally consequential respect. Ordinarily, we divide time into past, present, and future, taking the here and now for ourselves as the reference point. In Hume's words, we "imagine our ancestors to be, in a manner, mounted above us, and our posterity to lie below us."[5] Nothing is wrong in general with time seeming to be a succession of my todays leading gradually out of my past and into my future. It is difficult enough to get out of bed in the morning when one's focus is simply on the day ahead. If one also always needed immediately to confront the ups and downs of the past as well as the likely ups and downs of the future, it might seem, or indeed be, overwhelming. The neat conventional divisions into a long past and an indefinite future, separated by a manageably short present, is often helpful and for many purposes perfectly appropriate.

To some degree, we understand, however, that the segregation of our consciousness into present, past, and future is both a fiction and an oddly self-referential framework; your present was part of your mother's future, and your children's past will be in part your present. Again, nothing is generally wrong with structuring our consciousness of time in this conventional manner, and it often works well enough. In the case of climate change, however, the sharp division of time into past, present, and future has been desperately misleading and has, most importantly, hidden from view the extent of the responsibility of those of us alive now.[6] The narrowing of our consciousness of time smooths the way to divorcing ourselves from responsibility for developments in the past and the future with which our lives are in fact deeply intertwined. In the climate case, it is not that we face the facts but then deny our responsibility. It is that the realities are obscured from view by the partitioning of time, and so questions of responsibility toward the past and future do not arise naturally. Other times seem distant, and the people who then lived or will live in them appear to be irrelevant strangers. Acknowledgment of responsibility rests on recognition of connection. The climate connections are often not obvious.

Chapters to come will explore more fully some of the deep continuities inherent in climate change, but one obvious fact is the enormously long lead time built into some of the causal connections within climate. Carbon emissions injected into the atmosphere in a given year can contribute to forcing sea-level rise in not merely later centuries, but later millennia, dozens of centuries after the source of those emissions has disappeared from the earth. Some carbon emissions released early in the Industrial Revolution are yet to have their full effect, which still lies in the future. Present and future emissions matter as much as they do only because of past emissions and their long-lasting effects stretching far beyond the date of their release and on through the present. These long-lived connections provide a radically different example of the insight about psychology and culture from one of the characters created by my fellow Southerner William Faulkner: "The past is never dead—it is not even past."[7]

And similarly long chains reach from the present into the future. Conventionally, we tend to think that the future is yet to be born or is even only just beginning to be conceived. But the climate future was already beginning to take shape when humans started centuries ago to inject more carbon into the atmosphere than the usual climate dynamics could handle in the usual ways, and climate parameters were forced to start changing. The vast and accelerating carbon emissions of the late twentieth century and the early twenty-first century are building minimum floors under the extent of climate change in future centuries, barring radically innovative corrections of kinds that may or may not be possible and that we will discuss more fully in chapter 4. "The modes of common life that have arisen largely within the last one hundred years, and whose intensity has accelerated only since 1945, are shaping the planet for the next one thousand years, and perhaps the next 50,000."[8] The future is not inaccessible—we hold its fundamental parameters in our hands, and we are shaping them now. In this respect, the future is not unborn—it is not even future.

"The evil that men do lives after them, the good is oft interred with their bones," declared Shakespeare's character Mark Antony.[9] As an old man who, on the probabilities, ought to die fairly soon himself, I take considerable comfort in the knowledge that this dark assertion is an overbroad and skewed generalization (spoken as part of Mark Antony's maneuvers at Caesar's highly political funeral). The reach of the present—what we who are alive can set into motion—extends far across time for good as well as evil. In some cases—climate change is one—our reach will be long and deep, millennial and profound, whether we wish it or not. And we can make its outcomes good—or, at worst, far better than they would have been had we continued as we are headed now. One does better to heed what seems more likely to be Shakespeare's own voice and "to love that well which thou must leave ere long."[10] The generation alive now is the pivotal generation in human history with regard to climate change, because of three features of our historical context. And our responsibilities are awesome especially because of the implications of the third feature. We can first glance at the context and then begin to explore the grounds of responsibility.

The Context That Makes Us Pivotal

First, previous generations of humans have for around two centuries been changing our climate unintentionally and have left us with a global energy regime that now profoundly, progressively, and systematically forces the climate to change. The massive emissions of greenhouse gases (GHGs) that have resulted from the Industrial Revolution (and from the changes in the use of land, such as deforestation and the draining of wetlands, produced by the industrialization of agriculture) are disrupting the climate to which we and other living species had adapted over the previous ten thousand years of the Holocene.

Second, we are the first humans to understand the essential dynamics of our planet's climate; consequently, we have become aware of humanity's unintended subversion of its own environment through its uninformed past choices of energy sources. Scientists whose work is relevant to climate have produced remarkable—sometimes stunning—results. Much uncertainty remains, of course, but the basic outlines of climate science are clear—and far more advanced than they were only a few decades ago.[11] This impressive new knowledge puts us for the first time in a position to affect the climate intentionally by escaping from our inherited energy regime and to act on transition plans that have a reasonable chance of accomplishing their goals. Humans have been and still are in fact radically changing the planet's climate without any plan, but we are newly in the position to try to get a grip on the effects of our own behavior and attempt to exercise some intentional control.

Third, specifically what the science shows is that the default outcome is that the situation will become progressively worse, and human economic business-as-usual will make the future more threatening than the past for most living species—certainly including many humans, especially those with the fewest resources with which to adapt to the rapid, interacting changes.[12] Our unintended changes are rapidly undermining our own security. Therefore, we are not only the first to be able to understand what to do, but—most importantly—we may also be the last to be in a position to act before we exacerbate some major threats. This gives us an awesome responsibility. Humans have accidentally set our own house on fire, and if we do not douse the flames while they are no more extensive than they are now, it may not be possible ever to extinguish them. It is urgent for humans to get a grip on what we in aggregate are doing to the planet on which we live by blindly continuing the combustion of fossil fuels (and the destruction of natural ecosystems by industrial agribusiness) and instead to employ our recently gained knowledge of the climate system to design a transition to an energy regime that does not undermine our civilization.

Annual global carbon emissions in 2019 were the highest ever,[13] after a quarter century of mostly empty talk about tackling climate change, and the long-term trajectory of carbon emissions is at present sharply upward. Accordingly, the cumulative atmospheric accumulation of CO_2 reached its highest point in human history of 421.21 ppm in April 2021.[14] If other greenhouse gases like methane are also counted, the atmospheric accumulation was already in December 2020 the equivalent of about 500 ppm of CO_2.[15] The pre-Industrial-Revolution level was around 270 ppm, so doubling is well within sight. The explanation of why inaction would see matters worsen, and action is therefore urgent, is empirical and draws on various aspects of science. It is briefly summarized in the first chapter of a special report of the Intergovernmental Panel on Climate Change, *Global Warming of 1.5 °C,* usually referred to informally as the "Special Report on 1.5."[16] Here is the short version: because climate change is primarily driven by the cumulative atmospheric concentration of carbon dioxide (CO_2), and CO_2 that reaches the atmosphere is extraordinarily persistent, climate change will not stop becoming more severe until injections of CO_2 into the atmosphere completely stop—that is, until human society reaches net zero carbon. For any given degree of climate change, there is a budget of cumulative atmospheric carbon: more carbon, more change. To limit average global temperature rise to 1.5°C, annual global net anthropogenic CO_2 emissions must decline by about 45% from 2010 levels by 2030 and reach net zero around 2050; to limit the rise to below 2°C, CO_2 emissions must decline by about 25% by 2030 and reach net zero around 2070.[17]

If one follows the science, one can see why carbon emissions must rapidly be brought to net zero globally if future generations are to live securely. Every society's energy system needs to be completely decarbonized by totally eliminating the use of fossil fuels in order to stop the accumulation of carbon in the atmosphere within a relatively tolerable cumulative carbon budget. The *minimum* necessary for the safety of future generations, then, is a prompt global Energy Revolution. We are the pivotal generation, the only ones who can set the revolution strongly into motion while there is still time.

What I want to begin to explore a bit here, and elaborate in later chapters, is why we current humans ought to take the actions that are urgently necessary to stop climate change from becoming increasingly dangerous. The philosophically uninteresting reasons are self-interested, and there are tons of those. For example, the kind of megawildfires recently experienced in California and Australia as a result of climate-change-induced drought and heat produced horrifying human deaths and misery and monumental economic costs, including the contentious bankruptcy of the largest California utility, PG&E, which in turn threatens important renewable energy firms that were counting on long-term contracts with PG&E, which in turn undermines the firms' efforts to stay in business and provide energy without producing the damaging carbon emissions that contributed to the conditions for the wildfires—one downward social cascade.[18] We obviously need to protect ourselves from such economic vicious circles that undermine our own current interests.

Arbitrary Demands?

The focus in this book will instead be the converging moral reasons for action, which turn out to be multiple. I will concentrate on why we ought to act urgently and vigorously enough to protect future people from what will otherwise be the effects of perpetuating our present way of life permeated by an energy regime dominated by fossil fuels—why we ought to change to a way of life fit for a future on the carbon-sensitive planet we happen to live on.[19] The science demonstrates that if robust and extensive action is not taken, conditions for living things will be progressively more challenging and threatening. The remaining practical question is, why should we now have to do more than the very substantial amount that is already in our own interest? Climate change will worsen until it is limited, but a global Energy Revolution in the next couple of decades sounds like a heavy lift. Why should we in the here and now be expected to do so much? Does this not seem arbitrarily demanding?

For a start, that particular reaction seems at least as arbitrary as the situation might seem. One can focus on oneself and feel sorry for oneself: why poor me? But short-term and narrow self-interestedness is not inevitable, however easy, and it is no less reasonable to have a broader focus, on the rest of the planet besides oneself and on time beyond the immediate future, and to embrace the situation as an exciting opportunity to lead a meaningful and valuable life that could benefit people in the future—perhaps even all future people—or at least avoid depriving them of good options. How should one feel? Which attitude should one adopt? Should one resent the "burden" or embrace the "challenge"? Does the task demand inordinate sacrifice, or is it a historic opportunity to do something exceptionally worthwhile?

Our historical circumstances open up an intriguing prospect: what if some generations are called upon to meet challenges, or even to make sacrifices, that are unique to them? That do not fit any standard formula? Could this be? Climate scientists are telling us that we are now moving through an utterly crucial juncture. For a century and a half, carbon emissions were steadily climbing. Then, for the last three decades, they have been soaring: more than half of all the emissions since 1850 have occurred since 1986.[20] Now we must quickly make emissions level off and then "bend the emissions curve downward"—launch a steady decline in emissions at an angle across time that will bring the world to net zero carbon in two or three decades—certainly within the middle years of the lives of those who are now under forty years old.

Otherwise, the only ways to hold total cumulative emissions to any total compatible with any remotely tolerable amount of temperature rise, and of all the other manifestations of climate change, would be either a later precipitous plunge in carbon emissions that is probably politically and economically impossible and would be utterly catastrophic socially if it actually occurred; or miraculous amounts of carbon capture and storage (CCS), carbon dioxide removal (CDR), or solar radiation management (SRM), technologies in which the fossil-fuel companies have for decades steadfastly refused to invest, preferring to pour capital into exploration and production of more and more of the fuel that must never be burned without some such technologies. The choices, then, are in fact only four: dangerous levels of climate change from too much CO_2 accumulated in the atmosphere; a steady and sharp decline in emissions starting immediately; an unmanageable collapse in emissions later; or infeasible levels of CCS, CDR, SRM, or other geo-engineering technologies.[21] I am obviously omitting the shading in the picture, but this is the general outline.

Any way you slice it, it is absolutely crucial what the current generation does now. This is it. We face what in another context Martin Luther King Jr. called "the fierce urgency of Now."[22] But this may seem to ask too much of the current generation, each member of which is living the only life she will ever live and has some rights to enjoy. How, in the Churchillian phrase, could so much be asked of so few? Isn't the burden of promptly and firmly initiating a global Energy Revolution too heavy for the current generation alone to be expected to bear?

Unique Historical Period, Incomparable Moral Responsibility

This question about the perhaps exceptional extent of this generation's responsibility is difficult to consider in a sensible manner, and I explore this topic further in the second half of chapter 3 and in chapter 4. As people ordinarily think of generations, at any given time three generations are alive: grandparents, parents, and children. From here on, I refer to these three together, however, simply as "the current generation"—those alive now. The heart of one implicit complaint seems to be that it is somehow unfair to the current generation that the challenges we face are so much greater than what one might think is "the average burden for the average generation"—it's altogether too much to ask.

That, however, is an oddly ahistorical way of thinking, a bit like asking why I couldn't have been born into some other, pleasanter century. Perhaps the allusion to Churchill provides a hint. Was it fair that the so-called greatest generation of the 1940s had to confront the Nazis by themselves? Wouldn't it have been fairer if the task could have been shared with, say, the people of the laid-back 1960s (when I was "military age")? But the people of the 1960s could have helped to defeat the Nazis only if the Nazis were still undefeated in the 1960s, presumably by then much more entrenched. It is a good thing for all the rest of us who have followed that the generation of the 1940s rose fully to the occasion, and we remember them proudly.

Decades and centuries are not standardized, and we have no reason to expect the challenges they bring to be comparable. While reasonable sacrifices by individuals certainly have some limit, that limit seems to have nothing to do with any notion of standard generational burdens, a notion that could only be made to seem plausible by ignoring historical context. A complaint that burdens are unfair makes sense only if those burdens could be redistributed and thereby made fairer, but most large-scale historical challenges cannot be postponed, rescheduled for a more convenient time, or subdivided among different generations. Of course, some threats, if ignored, die away on their own. Other threats, if not met when their outcome is still up for grabs, build momentum or become entrenched and create radically new realities on the ground. Defeating them later may be harder than defeating them now, if it is even still possible. Contingent facts matter, and one must choose one's fights knowledgeably.

For better or for worse, you live at the time when you live. You confront what you confront. "You are here," as the street maps say. You can embrace your historical location or curse it—you are free in your response—but this response will have the effects that such a response has in such circumstances, at this point in history. You choose your response, but history, made by the earlier responses of others to their own circumstances, provides your social circumstances, which interact with your response to produce your effects and the different social circumstances to be faced by future generations.

We youths of the 1960s could not help with the battles of the 1940s because one cannot reach backward through history. The Hubble Telescope can show us what happened millions of years ago, but it does not enable us to intervene retroactively and change the universe's course. For purposes of action, time—for us humans, history—moves only forward. A complaint about generational unfairness in the case of historically embedded current challenges fundamentally makes no sense: no lighter schedule of imminent tasks is possible. The only alternatives we have are the responses actually available now. Historical context matters.

Moreover, while one cannot reach back to effect change, one can reach forward. In fact, one cannot avoid reaching forward! It is not that we could decide, if we wished, to stretch out and change future history, as if "history itself" were somehow already going to go in one direction of its own until people bent around and diverted it into a different direction from the one originally set. Unavoidably, we are partly making the future, like it or not: the human future, the planet's future, will head in the direction we set.

Historical context matters, but so do human action and choice. More precisely, future people's starting place will be where we leave off. The future is partly in the present. History is a continuing drama with narrative threads running through many generations, and we humans are powerful players in the drama. If we leave a planet with a climate still dominated by a fossil-fuel regime, the next generations will have to struggle to escape from this regime within the far worse climate that continuous combustion of fossil fuels will have produced by then, because we did not arrange a timely escape. Future people cannot reduce our challenges, but we—and only we—can reduce theirs. Or, we can indulge existing consumption habits and energy practices to make their challenges worse—our choice, their inheritance.

In the remainder of this introductory chapter, I want to separate out three closely connected reasons why it is especially important that our generation makes an exceptionally robust effort on climate change: (1) future generations will very likely face burdens and dangers greater than ours, (2) the worsening dangers are currently unlimited, and (3) less effort by us may well allow climate change to pass critical tipping points. In sum, the burdens and dangers for future generations will probably be worse and are now worsening, are still without limit, and are potentially unbearable.

That is why the struggle to stop the destabilization of the climate is our generation's fight.

Heavier Burdens

In the past, when philosophers and economists have thought about principles of intergenerational justice, they have usually assumed that there must be some kind of standard formula—for example, a single discount rate—that can be applied reiteratively. John Rawls, for example, originally wrote that when we are considering what principle should guide the current generation's relation to the next, we should ask what principle we could have expected the previous generation to have adopted with regard to us—a kind of slightly asymmetric intergenerational Golden Rule.[23] The previous people should have done such-and-such for the current people, and the current people should do about the same, but perhaps a bit more or a bit less, for the next people.[24]

Rawls's approach seems to me unhelpful for climate change because of three of its assumptions. The first is the one criticized in the previous section: that one can adequately characterize the situation faced by every generation in general terms in the abstract as if history had no integral fabric with periods of war and periods of peace, times of stability and times of disruption. Different generations may confront radically different circumstances demanding incommensurable types and levels of burdens and opportunities. His comment that "presumably this rate changes depending upon the state of society" does not seem to capture with sufficient vividness how extreme the differences between even adjacent generations can be in the wake of some exponential change.[25] Ideal theory cannot guide us.

Second, he formulates the obligations as exclusively between adjacent generations (with reiterations), without serious examination of possible direct obligations now to whoever lives in the distant future, whose situation needs to be taken seriously by current planning in the context of a phenomenon like climate change that locks in effects over millennia. The time lags between cause and effect in the dynamics of the climate system are far longer than most causal connections we ordinarily encounter.

The third is that what Rawls sees as called for on behalf of future people are mainly positive contributions: how much should we save for them? However, our current effects with regard to climate change on whichever people live in future is in fact quite different and potentially much more negative. All decisions about the degree of ambition for emissions mitigation in the present are unavoidably also decisions about how to distribute risks and burdens forward among this generation and multiple future generations, as we will see more clearly in the section on "Bequeathing Risks" in chapter 4. The less risk we bear, the more risk others bear. The avoidance of damage—and protection against damage—looms larger than the provision of savings, although the provision of alternative energy technology, which is a kind of embodied savings, is a crucial element in avoiding damage and disruption.[26]

Mitigation is more ambitious insofar as it contributes to reaching zero emissions of CO_2 globally at an earlier date—and therefore at a lower level of cumulative atmospheric accumulation of CO_2 and resultant climate change. That the extent of our efforts profoundly affects future climate dangers is fairly obvious. Even if there were one fixed quantity of risks that had to be dealt with by some combination of people now and people later, the fewer the risks that were dealt with by people now, the more of those same tasks would remain to be dealt with by people later, along with the tasks that would in any case arise only later. Of course, tasks for which now is the last chance may simply go forever unfulfilled.

But since over time the climate risks specifically are in fact expanding in number, increasing in severity, and in some cases feeding upon each other, that fewer of the present tasks are tackled means not only that relatively more of them may remain to be tackled in the future, if it is not then already too late, but also that the number and seriousness of the dangers will be absolutely greater than if we had acted decisively, because our failure to deal with current threats will leave open doors to danger that we could have closed. Not only may some of the original risks remain still to be tackled, but additional risks that could earlier have been headed off entirely will instead have emerged and need to be confronted too. Positive feedbacks cause some climate risks to reinforce others. For instance, when the white Arctic sea ice melts, the dark ocean water uncovered absorbs more heat than the white ice used to, and so the warming water melts the remaining sea ice faster still, revealing even more dark water. This is the primary reason the Arctic has warmed far faster than the remainder of the planet (which scientists call "polar enhancement" of temperature rise). I will look at feedbacks later in this chapter. For now, let us simply note that unless climate change is stopped, it will grow worse because, for one thing, in crucial respects it feeds on itself.

If we allow climate change to grow by unpredictable increments of severity over an unspecified further expanse of time, the burdens that fall upon some future generations may vastly exceed the burdens that we face now. And they will become heavier then than they would have been if we had acted more energetically now. Thus, the threats to whoever lives in future generations will be more severe by two distinguishable standards: more serious than ours are now and more serious than the burdens then would have become if we had acted otherwise. Is it fair to leave future people to face a much harder, and increasingly hard, challenge because we refused to face an easier one?

These heavier climate burdens also come in two distinguishable varieties: the biophysical and the sociopolitical. On the one hand, as long as CO_2 continues to be emitted by human economic activities, physical climate change will become progressively worse because it is the long-lived cumulative concentration of CO_2 in the atmosphere that is the primary driver of climate change. (1) The more cumulative CO_2 in the atmosphere, the greater climate change. In addition, climate change feeds on itself through positive feedbacks. (2) The more climate change, the greater climate change still. These are physical sources of heavier burdens for future humans for as long as the carbon emissions that drive climate change are not brought under control.

On the other hand, the current sociopolitical situation in at least the nations with the wealth and power—not in Yemen, not in Syria, not in Eastern Congo, but in most affluent countries—is relatively malleable, or at least not desperate, compared to the situation that is likely to arise in the future if climate change is permitted to worsen unabated. The current situation is unfavorable to climate action in important respects; for example, fossil-fuel interests now control the legislative branch of the US federal government and the Russian, Saudi, Australian, and Brazilian governments. The United States in particular confronts migrants fleeing oppression and poverty in Central America that is exacerbated by droughts and hurricanes worsened by climate change. US social discourse is increasingly uncivil and bitterly partisan, repeated local massacres are carried out with unregulated guns including military weapons, and Black citizens are regularly murdered by White police. Nevertheless, major nonviolent social change, including radical change in the energy system, still appears to be possible and recently to have become likely, although extremist violence is also growing. Neither the sociopolitical situation, which I discuss more fully in chapter 5 in the context of what action to take next, nor the physical climatic situation is yet completely out of control or impossible to change through essentially normal political action.

At some unpredictable time, if climate change continues to worsen, some phenomenon such as genuinely massive migrant flows—external or internal, from flooded coastal cities—may come to seem so threatening that the prospects for civil, cooperative, constructive responses are likely to decline much further and perhaps give way to social conflict. These are highly complex social phenomena, and I do not want to speculate. The fact is simply that, however unfavorable to cooperative action on climate change one thinks the political situation is now, politics could become much more dysfunctional, with social disruptions that would further smooth the path for demagogues with pseudosolutions and nationalists blind to global solutions, and such domestic deterioration could create higher obstacles to constructive international cooperation on problems that cannot be dealt with by individual societies. If we cannot accomplish positive multilateral action on climate change in our present situation, however absolutely good or bad one thinks it is, there is good reason to believe that positive action could become relatively much harder precisely when the physical/climatic threats themselves worsen. Future people, in sum, could face worse climate threats in less favorable sociopolitical circumstances with more violent and misinformed opposition. The challenges for future people will be greater than ours almost no matter what. The challenges will be greater still than they would have been if we do not do what urgently needs to be done now.

Unlimited Threats

More telling still is the crucial fact that, so far, we have failed to place any outer limit on the severity of climate change, either its physical manifestations or its sociopolitical effects. For now, physical climate change can simply worsen indefinitely. If certain tasks must be done sooner or later, and if we complete fewer of them, other people in future will need to do more. But climate change is not so benign. It is not merely that we may leave work against the dangers unfinished—the dangers are multiplying, thanks to our continuing profligate combustion of fossil fuel. If we do less now, the worst in future will not only be worse than it otherwise would have been, but, as of now, worse without constraint. Specifically, until CO_2 emissions reach net zero, there will be no limit on how severe climate change must become. It is intolerable that we should acquiesce in contributing to a potential runaway global danger.

If any human duty is unconditional, it is the duty to preserve the fundamental conditions, including the physical preconditions, of human society by avoiding dangerous threats to those conditions. Yet in blindly toying with the climate, we are daring to experiment with modifying these very preconditions of human physical and social life, in a manner famously noted well over half a century ago by Roger Revelle and Hans E. Suess: "Human beings are now carrying out a large scale geophysical experiment of a kind that could not have happened in the past nor be reproduced in the future. Within a few centuries we are returning to the atmosphere and oceans the concentrated organic carbon stored in sedimentary rocks over hundreds of millions of years."[27] Equally famously, and more sardonically, another great climate scientist, Wallace Broecker, quipped, "The climate system is an angry beast and we are poking it with sticks."[28]

No responsible scientist believes that climate change is yet fixed on a trajectory toward human extinction. But numerous scientists have embraced the idea that we should think of the period of history that we have recently entered as the "Anthropocene" because this has become an age in which the most powerful force changing our planet is aggregate human activity, including centrally the anthropogenic emissions that are increasingly modifying the planet's climate.[29] Without intending to, we are gradually wresting control of the climate, among other things, from the natural forces that used to determine it. Like someone who knocks the rider out of the saddle of a galloping horse and climbs on without knowing how to ride, we are taking control away without gaining control ourselves or, chillingly, having any good plan about how to keep it. The planet's climate is being thrown into confusion by the originally unpredicted and unintended effects of growing human consumption powered by expanding carbon energy.

The most elementary advice given to people who planned to visit a casino in the days before ubiquitous credit cards and phone banking was to decide while they were still at home how much was the maximum amount they could afford to lose and take only that much cash with them. In other words, put a firm limit on maximum losses in the face of uncertainty. Those who opt for less ambitious mitigation than is readily possible are ignoring this basic advice with regard to the well-being of future generations. They are leaving potential absolute losses for people in the future unbounded.

It is well established that as long as the atmospheric concentration of greenhouse gases—and especially CO_2—continues to expand, climate change will continue to become more severe.[30] And as long as CO_2 is emitted in amounts that produce net additions to the atmospheric concentration, that concentration will of course continue to expand. Accordingly, until global carbon emissions reach net zero, no outer limit on the maximum severity of climate change has been set. The severity of climate change can worsen indefinitely until carbon emissions reach zero. Decarbonization must be thorough and prompt to cap the atmospheric concentration, which requires vastly more ambitious mitigation than nations are currently committed to. The less ambitious mitigation is, the later the date that the atmospheric accumulation will stabilize. As long as climate change remains unbounded, the costs of the gamble inflicted upon future generations has no upper limit. This imposed gamble is explored more fully in "Bequeathing Risks" in chapter 4.

I should be more explicit about what I mean here by "unlimited," by which I am not claiming that the climate will change an infinite amount. And by "limited" climate change I do not mean a climate that completely stops changing and never changes again. As some opponents of action to slow the current climate change like to point out, the climate has always changed and will always change. The climate of the earth will always be influenced by changes in the earth's orientation toward the sun, and many other nonhuman factors.[31] I restrict "limited" and "unlimited" to anthropogenic change. Unlimited anthropogenic climate change is the maximum climate change that humans can cause. Limited climate change is less change than humans could have caused. So when I say that climate change is currently unlimited, I mean that nothing currently stands in the way of anthropogenic climate change becoming maximum anthropogenic climate change. We are currently on course to do however much damage to the climate of our own planet humans are capable of doing. But "anthropogenic" is a long word, so I will not keep repeating this qualifier. The sociopolitical downside also remains unlimited. This is chilling. Unbounded danger is difficult to judge judiciously. Human beings at their best are inexpressibly remarkable, with their indomitable spirits and their unrelenting resilience. I do not recommend generally betting against the human race. And yet, human civilization can be surprisingly fragile. Remember the armed looters in the British Virgin Islands and St. Martin after Hurricane Irma in 2017, stealing food and water from their neighbors, as often happens after disasters. Remember how quickly the Hungarian government of authoritarian Victor Orbán fenced out all Syrian refugees. Remember the rabid insurrectionists at the US Capitol in 2021, beating fallen police with flag poles. Somewhere in the shadows of stress, the social norms begin to tear.

That physical stresses lead to conflicting political demands, which themselves can rend the social fabric, is hardly a new insight. One may have supposed that Thomas Hobbes was displaying a capacity for dystopian imagination when he wrote, "There is no place for industry, because the fruit thereof is uncertain, and consequently no culture of the earth; no navigation, nor use of the commodities that may be imported by sea; no commodious building; . . . no arts; no letters; no society. And, which is worst of all, continual fear and danger of violent death; and the life of man, solitary, poor, nasty, brutish and short."[32] But according to Geoffrey Parker's monumental global history, *Global Crisis: War, Climate Change & Catastrophe in the Seventeenth Century*, Hobbes actually needed only to look around at the state of the world in what we now have understood as the heart of "the Little Ice Age"—roughly, the 1640s to the 1690s.[33]

The Little Ice Age consisted of climate change of only a single degree of average global temperature—downward, not upward, of course—but this modest bit of climate change, and especially the resultant disturbances to agricultural production and food prices, were one side of what Parker aptly calls a "fatal synergy" that was an exacerbating factor in a global mélange of troubles, ranging from the Thirty Years War in Europe to the violent Ming/Qing transition in China. Well before large numbers of individual people will collapse from heat stress from climate heating in the twenty-first century, their societies will be liable, at some unpredictable point, to become incapable of farsightedness, fairness, or even cooperation, and to disintegrate into conflicts over places to live and places to grow food, and over priorities for the distribution of these places.[34]

Yale historian Timothy Snyder has marked one more recent pathway steeply downward:

> When an apocalypse is on the horizon, waiting for scientific solutions seems senseless, struggle seems natural, and demagogues of blood and soil come to the fore. A sound policy for our world, then, would be one that keeps the fear of planetary catastrophe as far away as possible. This means accepting the autonomy of science from politics, and making the political choice to support the pertinent kinds of science that will allow conventional politics to proceed. . . . As Hitler demonstrated during the Great Depression, humans are able to portray a looming crisis in such a way as to justify drastic measures in the present.[35]

In the sociopolitical arena too, it is wise to call a halt well short of any cliff edges.

Let me be explicit about what I am not suggesting. In recent decades, it has become almost a reflex among moral philosophers to assert that climate action is urgent because otherwise the apocalypse is around the corner. I do not think that the apocalypse is around the corner—nor is human extinction, nor even—just yet—is the Hobbesian unraveling of civilization that I am invoking.[36] For now, I am appealing only to the solid fact that all such threatening possibilities (and many others less serious than these, but still serious) persist until we stop feeding climate change. Each kind of disaster is possible and can be reached easily from the route that we are now on, until limits make it impossible. A climate that is worsening indefinitely leaves all bad options open. We have inadvertently opened the barn door, and some of the horses have bolted. But other valuable horses, for reasons of their own, have lingered in the barn. For now, nothing is stopping them from leaving too. It is past time, then, to relock the barn door securely and save all that can still be saved.[37]

It would be reprehensible to take no action even if indefinitely worsening climate change were a purely natural phenomenon. (If it were purely natural, it would be more difficult to figure out how to go about slowing it down.) Yet, for us simply to sit idly by and watch anthropogenic climate change become progressively worse and prepare to engulf future people would be even more shameful and pathetically weak. And since we are in fact driving climate change with our own GHGs, especially our carbon emissions, we do at least understand what needs to be done for us to stop making conditions worse by our own behavior. And it is obviously all the more our responsibility—a negative one simply not to wreak havoc—for its being driven by our own actions.

Scientists have clearly explained various mechanisms by which climate change could escalate in severity, and empirical findings show that we are provoking, or coming near to provoking, particular ones of these mechanisms. I have previously tried to establish that if one understands the mechanisms, and one is finding evidence that one of them is being activated, that is all one needs to know for action to be required.[38] Lauren Hartzell-Nichols has developed this kind of argument extensively in her book *A Climate of Risk*.[39] One strength of this argument is that it has relatively weak premises: only the theoretical claim that mechanisms are understood, and the empirical claim that evidence is accumulating that the mechanisms are being engaged.

When one adds to the satisfaction of those two premises the fact that no limit has been established on how dangerous climate change may become,[40] it is utterly irresponsible for those of us alive now not to do our utmost to limit, at the very least, our own contributions to the future danger and instead to continue the present practices—most notably, casually burning vast quantities of fossil fuel—that are adding continuously to the mushrooming danger. On the one hand, how much, if anything, the people alive at a given time ought to do positively with regard to whichever people are yet to come, numbers and identities of whom are of course unknown because neither numbers nor identities have yet been determined by events including our choices, is a contested matter. Ought we to try only to see that they are no worse off than we are?[41] Should we try to make their lives better than ours, with or without discounting? And so forth.

On the other hand, the additional consideration here is comparatively and absolutely less controversial: we ought not to continue to increase dangers to future people—potentially, all future people—without limit when we understand the mechanisms by which the dangers can increase and have evidence that the mechanisms are in motion. I cannot imagine a plausible moral view that would not embrace this imperative. To deny this negative imperative would accord zero value, worth, and respect to numberless future people. To reject it would constitute calmly contemplating the possible undermining of the necessary conditions for civilized societies and even the possible creation of the sufficient conditions for human extinction, since climate change can produce levels of heat and other phenomena that humans cannot endure.

Tipping Points

We have in addition another compelling reason for robust immediate action in recently acquired understanding of the dynamics of the planetary climate. While we are discussing climate change that is anthropogenic, it is crucial to keep in mind that whenever human action sets off positive feedbacks, what was originally anthropogenic to some degree takes on a life of its own.[42] Andreas Malm characterizes it nicely: "Society having touched off climate change, nature does the rest of the work. . . . Global warming is not built but triggered."[43] We are confronting climate change that is initiated by humans, but this increasingly includes the positive feedbacks produced by the already operative climate dynamics into which we are recklessly intruding. Extensive theoretical understanding and solid evidence suggest that we are in fact approaching a number of critical "tipping points," such as threshold ocean surface temperatures that will precipitate the collapse of massive Antarctic ice sheets and Greenland ice sheets, driving sea levels much higher and gradually inundating sea coasts around the world and driving populations out of cities with locations like Mumbai, Shanghai, Miami, and New York—and, of course, much of The Netherlands and Bangladesh.[44] While the current climate change is anthropogenic—driven by society's failure to mobilize against the primitive and dirty energy sources of coal, oil, and gas—the direct changes provoke natural responses that feed into further change. In short, while humans began the process, it could run away. Notoriously but horribly, natural species have been rapidly crashing for years.[45]

If enough positive feedbacks fed into each other to launch what the scientists call a cascade of positive feedbacks, it could lead to a "Hothouse Earth."[46] These scientists are now suggesting that a cascade is liable to begin soon if we persist in our failure to take prompt and serious measures to reduce carbon emissions. In my argument here, however, I pull back to the simple fact that for now cascades remain possible because climate change remains unlimited. I do not rely on an assumption that a cascade of positive feedbacks is definitely about to begin or even that it is likely to begin, although this may very well be true. I assume only that an anthropogenically launched cascade is entirely possible because various individual tipping points are likely to be passed as long as climate change continues without restraint.

It is, then, *likely* that the near future is the last chance to avoid passing significant tipping points and entirely *possible* that the near future is the last chance to avoid provoking a cascade of tipping points.[47] These tipping points are significant because they unleash either or both of two conceptually separable but often empirically inseparable processes. Passing a tipping point means triggering irreversible change, and frequently these irreversible changes themselves also become long-term positive feedbacks, sometimes exacerbating other processes of climate change so that they become exponential, a "cascade." Conceptually, a change's being irreversible and a change's being a positive feedback are two different matters. (And changes' being at any given time unlimited in number is a third matter.)

Obviously, even an irreversible change that was not a positive feedback would still by itself contribute to making climate change worse until it ran its course. If the melting of an ice sheet becomes irreversible, the melting will contribute to sea-level rise (because the water in ice sheets now rests on land) until the ice has all melted. When the ice is all gone, that particular process will stop. This melting may be irreversible, but the process need not continue indefinitely. So a particular process could be irreversible, but also be limited, and not be the source of positive feedback.[48]

Nevertheless, if one process after another will contribute to making climate change worse, the overall climate change would worsen without limit, even if each contributing process would run a limited course, until all contributing processes had run their various courses. If the West Antarctic Ice Sheet melts, that will make sea-level rise worse until that ice is gone. If the Greenland Ice Sheet melts, it in turn will make sea-level rise worse until that ice is gone. So too the East Antarctic Ice Sheet.[49] Each case of melting is obviously constrained by the amount of ice available to melt, but the quantities are staggering. Recent calculations show "that Earth lost 28 trillion tonnes of ice between 1994 and 2017. . . . The rate of ice loss has risen by 57% since the 1990s."[50] Climate change—here, sea-level rise—can continue until all the ice in all the ice sheets is gone. At some point, many millennia from now, the residual effects of humans might fade out, especially if humans are gone. Or perhaps the course of geological history would have been forever diverted from the path it would otherwise have followed if humans had not existed and created an Industrial Revolution based on moving carbon from under the earth into the air. Either way, the damage caused would be immeasurable.

Conclusion

I have separated out three conceptually distinct strands of the basis for urgent and robust action to stop climate change from worsening: inevitably more difficult challenges for future people, no limit yet on the extent to which humans will modify the climate, and the danger of passing critical points of no return: tipping points that launch irreversible change. What is truly scary is empirical combinations of two or more of these factors, especially if one of the factors is the third: passing tipping points for abrupt worsening. For instance, it is already worrying that we have so far imposed no limit on the disruption that we are causing to the climate, but that could mean only that we were very slowly and incrementally making matters worse for a while. But if we leave the disruption unlimited for long enough that we meanwhile pass critical tipping points like initiating irreversible melting of additional major ice sheets, then the most limited that the damage can possibly ever be will be far worse than otherwise. The other side of the coin, of course, is that if we throw ourselves into the effort, we can make a huge positive difference for the lives of virtually every future person who lives on this planet.

The processes sketched just above are concrete embodiments of the reality that time is continuous, not partitioned, except for practical convenience in our own consciousness. Any attempted separation of the flow of human history, its causes, its effects, and the responsibilities of those of us who will unavoidably contribute to the future direction of the flow into discrete periods is at best an oversimplification and often an illusion or an evasion. I live amid the wealth, ease, and technological wonders of a postindustrial society only because of the fossil-fuel combustion that drove the

Industrial Revolution in my past and created my present standard of living. My present immersion in a growth-obsessed, plastic-strangled consumerist society that still burns ever more fossil fuels each year (except the pandemic year of 2020) is locking in critical and dangerous features of the future climate of people I erringly tend to think of as distant strangers. For me to deny that this past and this future are part of who I am and what I do would be to fail to acknowledge fundamental realities and to shirk inescapable responsibilities. Or so I will try further to show in what follows.

2

The Presence of the Past

You know, they straightened out the Mississippi in places, to make room for houses and livable acreage. Occasionally the river floods these places. "Floods" is the word they use, but in fact it is not flooding; it is remembering. Remembering where it used to be.[1]

It might then be discovered that the agent has done something she had never dreamed of. She owns the action no less for that. . . . The outcomes . . . are integral aspects of the original action as stretched out over time. Global warming is an integral aspect of consuming fossil fuels, not *another* action performed by others. . . . The fact that humans act within the carbon cycle and other circuits of nature does not in any way diminish our agency. It amplifies it.[2]

For it is only by assuming full responsibility here for one's own elsewhere, only by assuming full responsibility today for one's own yesterday, only by this unqualified assumption of responsibility by the "I" for itself and for everything it ever was and did, does the "I" achieve continuity and thus identity with the self.[3]

Why is it so important that humans urgently cease injecting CO_2 into the atmosphere in the future? Present and future emissions matter primarily because we humans have already injected so much CO_2 into the atmosphere in the past. As mentioned in chapter 1, the earth's climate system has a single carbon budget for any given probability for any given amount of change in the foreseeable future. This means that the carbon budget is both global and cumulative, and that means: no separate budgets for past, present, and future. There is one carbon budget for all the centuries of any relevance to humans. The proportion of the budget that was used up in the past is therefore not available in the present or the future. That much of the budget has been spent; that much CO_2 has already been loosed upon the planet; and much of that CO_2 will remain in the atmosphere for significant stretches of the future, forcing the climate to change, unless the CO_2 could be actively removed and once again securely sequestered under the land or the sea from where fossil-fuel companies still persist in extracting it.[4]

Of course, the emitted CO_2 is not unchanging. CO_2 gradually decomposes by each of several different natural processes, which are complex and which scientists have only recently begun to grasp. Different processes operate over different time-spans, so different bits of CO_2 remain in the atmosphere, the oceans, the vegetation, and the soil for different lengths of time. The various gradual rates of natural decomposition and removal can be built into climate models, and scientists can probabilistically compute them more precisely.[5] But from the perspective of human life and history, the removal processes are all slow, and the slowest are very slow indeed. Natural processes have no chance whatsoever of removing CO_2 from the atmosphere at any rate remotely approximating the ferocious rate at which human activities are spewing it into the atmosphere. For all practical purposes—for all human decisions about policies affecting the climate—once the CO_2 is injected by combustion into the atmosphere, large portions stay there. In this case, what goes up mostly stays up. The earth has a single cumulative carbon budget for the human past, the present, and as much of the human future as we could conceivably foresee and care about—for multiple millennia.

This book is primarily forward-looking. Far more attention is given to the presence of the future than to the presence of the past. But we need first to take a brief look at the past and the ways in which it permeates the present in order to appreciate the stringency of the current requirements about what to do about climate change now; that is the task of this chapter. Contemporary climate change did not invade the earth from somewhere else in the universe. Climate change is a product of this planet, made here by particular choices by particular people in specific nations at specific times—most critically, of course, the choices that led to the Industrial Revolution fired by fossil fuel and to the unleashing of our current climate change by the carbon emissions from those fuels. A concrete human past created the concrete human present that is threatened by accelerating climate change.

At the conclusion of the Conference of the Parties to the UN Framework Convention on Climate Change in 2018 (COP 24), the US federal government issued the following stark declaration: "The United States is not taking on any burdens or financial pledges in support of the Paris Agreement and will not allow climate agreements to be used as a vehicle to redistribute wealth."[6] This official statement by the US Department of State implicitly assumes the nation-state as the unit of analysis. Within the human species, I take individual persons to be the units of ultimate value; I take their rights to be supremely important; and I take nation-states, or governments, sometimes to have instrumental value but never to have ultimate value. At their best, states are servants of their own people without being threats to the people of other states; states rarely perform so well. In this chapter, nevertheless, I want to confront this official position of the US federal government in its own nation-state terms, which are after all the terms of the Paris Agreement of 2015 and virtually all international climate negotiations, for better and for worse. For the sake of all the arguments in this chapter, I will accept the nation-state as the unit of analysis.[7] The last and most extensive of the three arguments in this chapter, built around the introduction of the concept of sovereign externalization, is a critique of (national) state sovereignty as it is normally practiced and as it is implicated in climate change.

Owning Our National Past

Throughout the international negotiations since the adoption in 1992 of the Framework Convention on Climate Change, the US federal government has insisted that past national emissions are not the basis for a present national responsibility for dealing with the climate change that those past emissions are now causing. Many other national governments acknowledge some degree of what is usually called "historical responsibility": present national responsibility that is basically proportional to past national contributions to the problem, measured ordinarily by cumulative past national emissions (or per capita cumulative emissions), which can reasonably be thought of as the proportion of the whole world's cumulative carbon budget consumed already by any one nation. In this chapter, I examine this issue of historical responsibility and explain why the above position adamantly asserted by the US federal government is indefensible and in denial of reality. Many state- and local-level US political leaders, and many national leaders throughout the world, have rejected the efforts of the US federal government to divorce itself from history in this way. I will try to show briefly why these critics of the federal government are correct.

This chapter presents three independent arguments for what I take to be the most intuitive understanding of historical responsibility: a pure fairness argument and two wrongful imposition arguments, one in which what are imposed are seen as violations of rights and one in which what are imposed are seen as ordinary economic costs. The persistent calls in international negotiations for historical responsibility for climate change are always calls for the acceptance of accountability for the full consequences of a historical process of industrialization that relied on burning fossil fuels and that occurred nation by nation, but imposed the consequences of that reliance on carbon combustion on humanity generally. However, not everyone means the same thing by "historical responsibility," and the three specific arguments presented here are not the usual ones found in the literature.[8]

The nations with historical responsibility are those who contributed to climate change by deriving the energy for their industrialization from the combustion of carbon-based fuels.[9] No nation has yet shouldered anywhere near the full consequences thereby created for human health and environmental stability across the world generally. Those problems include lethal air pollution, especially the particulate matter from burning coal that was present from the beginning of the Industrial Revolution in the grimy pioneering Welsh and English factory towns; ocean, river, and groundwater pollution from perennial oil spills; perpetual methane leaks from extraction and transport of gas; and, worst of all, CO_2 emissions that began to exhaust the climate system's limited capacity to process additional anthropogenic CO_2 without raising surface temperatures throughout the planet.

Greenhouse gases are multiple, but by far the most important greenhouse gas cumulatively to date is CO_2 from combustion of fossil fuel: fossil fuels "are, by far, the largest contributor to global climate change, accounting for over 75% of global GHG emissions and close to 90% of all carbon dioxide (CO_2) emissions."[10] The primary source of the anthropogenic CO_2 accumulated in the atmosphere has been the national processes of industrialization, which have occurred in different nations at different times, even different centuries. Great Britain industrialized in the eighteenth and nineteenth centuries, while India industrialized in the twentieth and twenty-first. The damage from industrialization has been universally distributed in the form of, among other things, the growing dangers constituting climate change that face everyone, including everyone in future generations of all nations, and the massive problems of air pollution that kill millions every year.[11] The contention of the national climate negotiators who acknowledge historical responsibility for climate change, then, is that the nations that were the first movers and unilaterally controlled their own processes of industrialization should restore the global playing field to a more level position by bearing their share of the burdens that are resulting from their disruption of the global climate with the still-accumulating greenhouse gases they have been injecting into the atmosphere. This contention is grounded in the unilateral national causal contributions to global harms and other global costs from the processes of national industrialization, including the industrialization of agriculture by agribusinesses.

In the aforementioned US State Department's declaration at the end of COP 24 in 2018, the US federal government presents itself as defending Americans against greedy claims for redistribution by foreigners who would take away what is rightfully ours. By contrast, climate negotiators advancing claims of historical responsibility are demanding that the United States and other highly industrialized nations take back upon themselves the dispersed costs of their own industrialization that have so far been dumped upon humanity generally. The US federal government insists, on the contrary, that the benefits of US industrialization are rightfully all the United States' own: all benefits are rightfully retained, while all costs (wrongful harms and ordinary costs) are rightfully scattered—and properly left to lie wherever they fall. In fact, industrialization has involved a massive and time-lagged—and therefore largely hidden—distribution of costs imposed across the planet and into, among others, countries that had not yet industrialized. This past dispersal of delayed costs is not past; the ongoing costs simply remained invisible until the climate scientists made the effects of the past carbon emissions on the future climate more apparent to eyes ordinarily focused on the present, and in many cases distracted by decades of lies about climate change and diversionary tactics on the part of the defenders of the energy-business-as-usual.[12]

The contribution to damage done begins with the role played in the creation of the global carbon energy regime that relies on extracting, transporting, refining, and burning of fossil fuels, whose emissions continue to force the climate to become increasingly inhospitable to human flourishing. The international carbon regime greatly benefited some societies for a time by fueling national industrial growth but is now harming practically all by undermining the stability of the climate and increasingly the stability of the very economy created.

Disowning Our National Past

To this contention that the nations that are the initiators and the proprietors of industrialization should mainly bear its costs, the primary objection has been that the contention is harsh—unfairly harsh—for two main reasons. Individuals in the present and future in the industrialized nations would, it is claimed, suffer for "crimes" they did not themselves commit and for "crimes" that were not crimes when the actions in question were done. Invoking historical responsibility is thought to be like passing an ex post facto law and then, since the perpetrators of the newly minted "crime" are dead, punishing their children and grandchildren for this "crime" that was anyway not a crime at the time—a double injustice. We would have targeted the wrong people even if there were an offense, and anyway the relevant action was not an offense when it was done: wrong person and no offense.

First objection: "wrong person." Even if there were an offense, the offenders would now be dead. It is not fair, the objector urges, for present and future individuals to suffer for the sins of past individuals. Why should the present and future generations of the causally responsible nations pay the debts of previous generations of their respective nations?

The nation, like many communities and collectives, is a historically continuous entity of which current individuals are members. The United States began in the eighteenth century and is still going. I could not have requested or consented to the carbon emissions of my US ancestors, of course, but I grew up and lived in the national economy they built through the processes of industrialization that caused those emissions. A nation contains continuing structures and institutions; past, present, and future members inherit these ongoing national formations and practices, including capital and infrastructure. We living stand to our predecessors in our own communities partly like someone who is the executor of a dead person's estate: we must pay any remaining debts. The living compatriots are the executors simply because there is no one else who can reasonably be expected to do it, on the standard view of nations assumed by the US State Department.[13]

History does not consist of hermetically sealed separate acts with completely unrelated plots and entirely new casts. Human affairs are, for better and for worse, a continuing drama with narrative threads that run through many generations, as in Faulkner's dictum: "The past is never dead—it is not even past." The objector is attempting to construct dams in the continuous flow of national history. But are there any liabilities for the present generation to inherit from past generations? By what standard did previous generations do anything for which accountability needs to be determined?

Second objection: "no offense." In most discussions of historical responsibility for climate change, the contribution to the creation of the problem is conventionally measured by cumulative emissions of CO_2 by individual nations, but the exact formulation of this measure requires more detailed discussion.[14] Any suggestion that past emitters, and now their descendants, are accountable for those emissions would, an objector might say, create a crime ex post facto, or perhaps impose carbon pricing ex post facto. It was not illegal or otherwise wrong simply to release the carbon emissions at the time of the Industrial Revolution and since, and certainly no requirement to purchase an emissions permit or to pay a carbon tax existed when the earlier emissions were released. So why should anyone be punished or charged simply because emissions were innocently released?

The answer to this objection is somewhat complicated and occupies the remainder of the chapter. I present two different kinds of argument that, I believe, are both valid, and both point in the same general direction for policy. The second kind is more critical of customary practices than the first, and it in turn branches in two directions. I believe that the two arguments of the second type also cut more deeply, and we will see in chapter 3 that they have stronger implications for policy. I will refer to them respectively as "the pure fairness argument" and "the wrongful imposition arguments." The pure fairness argument focuses on the clear and straightforward fact that various generations of the early industrializing nations have contributed to the problem of climate change by creating and developing the carbon energy regime that today continues to drive the worsening of the climate. The two wrongful imposition arguments consider the major significance of the source and extent of the relative benefit to those who brought about climate change, while the pure fairness argument leaves relative benefit aside.

The Pure Fairness Argument

We should readily grant that no individuals in earlier generations ought to have been *punished*, for there was indeed no crime, and we must also insist that all talk about crime and punishment rests on a bad and irrelevant analogy. An objection that assumes that someone is being said to be at fault would be denying a thesis that I am not advancing. What we now understand is that the Industrial Revolution was powered by an energy revolution that produced today's global energy regime based on the exploration for and the extraction, transport, and combustion of fossil fuels.

The combustion of fossil fuels is the main driver of climate change, although deforestation and other changes in land use and agricultural practices are also important. What is undeniable is that the combustion of fossil fuels is the main contributor to climate change. So it is accurate to say that the burning of gas, oil, and coal is causally responsible for most of the climate change.[15] What the objection "no crime and therefore no punishment" gets right is that causal responsibility does not entail moral responsibility.[16]

Moral responsibility is a complex matter, but insofar as it involves being morally blamable (and therefore perhaps deserving of punishment or criticism), what one caused must have been morally wrong. One must be at fault for causing what one caused. I may be causally responsible for a new tree appearing in the backyard if I planted the tree. But if there is nothing wrong with having another tree in the backyard, it would be misleading to say that I am morally responsible for the appearance of the tree (unless it is to thank me rather grandly for my gardening effort).

A proponent of the "no crime" objection in the case of the fossil-fuel regime would be using the misleading analogy of crime and punishment to acknowledge the obvious proposition that our ancestors are indeed causally responsible for the existence of the fossil-fuel regime and its vast carbon emissions. But, the objector would in effect be adding that, since it was not wrong to burn more coal, and then more oil and gas as well, it would be incorrect to say that our ancestors are morally responsible for the fossil-fuel regime and its effects. They are causally responsible, but since no wrong was done, there is no moral responsibility.

The pure fairness argument grants this: causal responsibility, yes; but *no moral responsibility, because no wrong specified* (or even alleged), and therefore no fault.[17] The fact remains that the accumulated effects of the fossil-fuel regime have long been undermining the climate and thereby undermining human economies and societies. Therefore, it is vital that humanity close down the fossil-fuel regime and make the transition to a safe energy regime—and quickly. A rapid transition to a sustainable energy regime will have substantial costs.[18] Energy-poor developing countries, for instance, must be enabled to the greatest extent possible to leap-frog over carbon-based energy to noncarbon energy, because their carbon emissions will exceed the remaining carbon budget if they follow the same high-polluting path originally taken by the now-developed countries. Instead of building more gas pipelines on the basis of the corporate myth that gas is a necessary "bridge," we need to build more nimble electricity grids with large storage capacities and artificial intelligence to manage flows of solar and wind energy.[19] Coal miners and their families who are being abandoned by the executives of the coal companies for which they faithfully worked, with their health care and pensions left unfunded and their communities polluted, while the coal executives award themselves obscene amounts of severance pay, ought to be cared for or retrained by the rest of us.[20] The manufacture of gasoline-burning cars must be replaced as fast as possible by the manufacture of cars that do not burn fossil fuel, and so on.

The unavoidable question is, who is going to pay for the urgent transition between unsafe and safe energy regimes? If the transition is to occur, someone must shoulder the costs. Jules Coleman put the general issue crisply: "Who bears losses in the absence of fault"?[21]

And the moral aspect of the question is, what is a fair allocation of these payments? On whom is it fair for these costs to fall? This is a question neither of retributive justice nor compensatory justice, but of distributive justice—basic fairness. Instead of an analogy of crime and punishment, consider a better analogy that acknowledges that all parties are faultless, although this analogy is inevitably still highly oversimplified.[22] Four of us need to walk across a small desert, and we each have one trunk. We can find only one camel, so we decide to load all four trunks on this camel. Unknown to us, this camel can only carry three trunks. Three of us place our trunks on the camel without incident, but when the fourth adds her trunk, the camel breaks down. Now none of us are able to make the trip.

Two observations about this little adventure, while literally true as far as they go, are deeply misleading, but a third is more insightful. The first observation is that the first three trunks loaded onto the camel produced no harm to the camel. The second observation is that it was the fourth trunk that caused the camel to break down. But each observation is so partial that it distorts. It is true that it was the fourth trunk that caused the camel to break down and that the camel would have been fine carrying the first three trunks. But the camel broke down because he was asked to carry four trunks, not the fourth trunk alone. The fourth trunk caused the camel to break down only because he was already bearing the first three. It may be that in one sense "the first three trunks loaded onto the camel produced no harm," but they prepared the way—created the conditions—for the harm to occur: they created the situation in which the fourth trunk would do damage by exhausting the camel's capability. It was four trunks that brought down the camel: the first, the second, and the third, as well as the fourth.

The further analogy of a budget may also be helpful here: the weight "budget" for using this camel without harm turned out to have been three trunks; while the first three trunks did not crush the camel, they exhausted the no-harm "budget." The weight of the first three trunks is why the camel was broken by the fourth. The fourth was the precipitating cause of the breakdown, but it is very far from the whole explanation.

The first three travelers whose trunks exhausted the camel's carrying capacity did nothing wrong when they placed their trunks on the camel's back. The story grants an assumption parallel to the objector's contention in the case of the energy regime and climate change that no one in the early generations of the Industrial Revolution was committing a crime by burning fossil fuels (leaving aside the deaths from the air pollution from coal burning, which was understood, but ignored, long before climate change was well understood). The first three travelers committed no crime, are without fault, and do not deserve any punishment. If they take responsibility jointly with the fourth traveler for making some satisfactory arrangement to deal with all four trunks, they are not confessing that they are villains in the story. They are simply acknowledging that they too are fully and irremovably part of the explanation of the outcome. The first three travelers "own" the outcome as much as the fourth. The explanation for why the fourth traveler has no way to move her trunk must include the first three travelers. Even if neither bad intent nor foresight was involved, the actions of the first three contributed crucially to the bad results. While not villains, the travelers are still accountable agents. And all four travelers together are accountable for breaking the camel's back even though they are equally faultless.

To some, it may seem unfair that the first three travelers should have to bear a burden because they performed a perfectly innocent act with no bad intent. That may seem unfair, but it is only unfortunate—and unfortunate for all concerned. The fourth traveler also performed a perfectly innocent act with no bad intent. What would be unfair is if the fourth traveler alone had to shoulder the loss of the camel. A world that contained stronger camels would have been a more fortunate world for desert travelers. A planet on which CO_2 did not block heat from escaping through our atmosphere would have been a more fortunate planet for humans who wanted to burn coal. But camels carry what they can, and the planet's atmospheric chemistry works the way it works.

The emissions/trunks placed on the climate/camel have not all been the same size. This is a further complication. It seems reasonable that everyone who is accountable for the bad outcome of accelerating climate change ought to share proportionally the costs of developing alternative energy so that we can stop burning coal, oil, and gas very, very soon. China's and India's present and future emissions are a problem only because of the United States' and the European Union's past (and present and future) emissions—and of course now increasingly China's and India's own past emissions.

The four different travelers are obviously the analogues of the various nations who in various time-periods created and expanded the fossil-fuel regime. Even if one assumes that every generation in each nation simply meant well and was trying to improve their own lives and the lives of their children, they each contributed to the creation of a profoundly dangerous problem: rapidly worsening climate change. They acted without fault, but they caused a severe threat that it will be expensive to control. The costs of an energy revolution away from fossil fuels must be paid. It is unfortunate that this is so, but if the costs are to be covered, some people must pay. On whom is it least unfair that this burden should fall? The camel story suggests that every nation whose earlier generations contributed to the problem should contribute to the solution in proportion to their contribution to the problem.

Why? Because this outcome is proportional to their actions and their choices. In this respect, it is their outcome—they "own" it even though they did not foresee it. Responsible agents acknowledge what they do, including the consequences they did not foresee and would not have chosen, as the epigraph from Václav Havel's letter from prison suggests. Other potential agents also may not have foreseen and would not have chosen the actual results, but the difference between those other agents and these agents is that these agents brought about these results. These are their results, and they are accountable for them even though they may not deserve to be blamed or criticized for them. To refuse even to acknowledge them, as the US federal government has refused to acknowledge the size of the Unites States' role in the history of the emergence of climate change, is deeply dishonest—an evasive attempt to disown consequences that one has in fact produced. Someone must unfortunately pay to deal with a problem that has been created. Industrializers have faultlessly made a mess—who ought to clean it up?

The issue is fairness, not blame. The four travelers in the camel story did nothing wrong in loading on what they did not realize were backbreaking trunks, and the early industrializers did nothing wrong in relying on a process that they did not know was combustion producing climate-undermining carbon emissions. Unforeseen difficulties arose from innocent actions; this happens all the time. The question then is, who is it fairest to expect to sort out the resulting difficulties—innocent noncontributors or innocent contributors? One could argue that since all are innocent, and no one is to blame, the choice of people to deal with the problem might as well be random: flip coins or hold a lottery.

That attitude, however, would best fit a disjointed world in which past, present, and future were not connected in meaningful ways, and each morning saw an entirely new world in which all options were equally open. For a random world, random procedures might seem appropriate—why not? However, in a world with continuities in which the present does not merely occur after the past, but flows out of it, and the future does not simply happen next after the present, but is partly shaped by present actions and choices, it seems far more appropriate—more in tune with reality—to acknowledge the continuities by accepting accountability for the actual results of what members of one's own community have in fact done. One can sometimes truthfully say, "I never meant this to happen," while acknowledging correctly that this happened because of what one did as a result of one's own choices. If "this" is a mess, simply to walk away from it is to act as if the connection between us and it did not exist or meant nothing. Walking away is evasive and dishonest because it does not face the fact of national connection and of national continuity with both contribution and outcome.

The pure fairness argument concludes, then, that since some individuals must pay to deal with climate change, it is fairer that they be the contemporary and future compatriots of those who created the carbon energy regime and climate change. Individuals are accountable for the bad results as well as the good results of their own nation's energy history. If a share in the national wealth generated belongs to them, the corresponding portion of the climate damage generated belongs to them too.

The Wrongful Imposition Arguments

We turn now from the results of the agency of individual members of one's own nation-state to the results of the agency of the corporate entity, the sovereign state, itself—the consequences of national policies. I will write mainly in terms of the policies of sovereign states, but it is worth bearing in mind that the world's largest fossil-fuel firms are all state-owned and in this respect part of a sovereign state. Saudi Aramco's policies are policies of the Saudi state, just as Gazprom's policies are policies of the Russian state, and Sinopec's are policies of the Chinese state. While sovereign states readily claim the wealth resulting from national industrialization, they do not readily claim the damage resulting. In fact, the early industrializing nations that today count as developed have clung to the wealth produced by their respective Industrial Revolutions, but some—most loudly and insistently for the quarter of a century of climate negotiations, the US federal government—have treated the problems produced as shared by all, beneficiaries and nonbeneficiaries alike. Climate change is treated as a global responsibility in spite of the fact that the bulk of the CO_2 emissions were until recently produced by the minority consisting of the now-wealthiest nations in the process of generating their own national wealth. In recent decades, China has become the largest source of emissions—and a far larger source than it needs to be.[23] When one retains most of the benefits from a process and sheds as much of the costs as possible, one's net benefits are greater than if one shoulders all the costs of one's own enrichment. And if others who bear some of those costs do not also receive at least the same amount of benefits as of costs, they are made absolutely worse off than they would have been if one had done nothing.

A firm's maximizing its gains by minimizing its costs through leaving all the environmental, health, and other problems it creates to be handled by society is a phenomenon that economists call "externalization." Since the term accurately indicates the movement of dismissal, ejection, or alienation, I will retain it but focus on different aspects of the process and suggest quite different reasons for condemning it.[24] So far as possible, a purely self-interested firm controlling a productive process owns the good products and disowns the bad products, such as pollution and disease, which become in economistic jargon "negative externalities." The selling price of the good product does not incorporate the cost of dealing with the negative externalities, because they are simply not dealt with, but are dumped onto society as a whole. This makes products cheaper to buy, easier to sell, and so usually more profitable. From the point of view of society as a whole, it is irrational to allow such externalization at the level of firms, because self-interested firms have no incentive to avoid endlessly creating environmental and health problems, which cost them nothing, from which others will suffer, and for which others will have to pay if and when the problems are cleaned up. A basic function of good government, even for advocates of small government, is to design incentives and regulations in such a way as to drive the internalization of the costs of negative externalities so that those who purchase a product pay a price that covers the prevention of, or the remedy for, the problems that would otherwise be created and dispersed by the product's creation and sale.

By exactly the same reasoning at a more general level, from the point of view of humanity as a whole, it is irrational to allow externalization at the level of nations, because purely self-interested nations—nations pursuing exclusively their own national interest— have no incentive to avoid creating environmental and health problems from which the rest of humanity will suffer and for which the rest will have to pay. I will call this "sovereign externalization," and the creation of global climate change, when combined with a refusal now to fund the initiatives necessary to stop it, is a monumental example of sovereign externalization in action. The current system of sovereign states encourages sovereign externalization— externalization at the level of the nation—and, in addition, makes it difficult to resist, even though from the perspective of humanity as a whole, especially future generations, allowing such negative sovereign externalities is wildly irrational.

Amitav Ghosh captures this picture beautifully: "That ultimate instance of discontinuity: the nation-state."[25] In the case of climate, the would-be externalization may also be self-defeating at the national level because the undermining of the global climate damages every nation and the world economy, on which national economies depend. Years ago, the eminent geophysical scientist David Archer put the fundamental point lucidly: economics "is a description of the way that money, the lifeblood of our economic system, really flows, analogous to the statement that water flows downhill. Money flows toward short-term gain, and toward over-exploitation of unregulated common resources. These tendencies are like the invisible hand of fate, guiding the hero in a Greek tragedy toward his inevitable doom. Our understanding of economics tells us that the free hand of the market, also known as business-as-usual, will not cope gracefully with the threat of global warming. Ultimately the question may come down to ethics, rather than economics."[26]

The earlier fairness argument already showed that it is unfair to refuse to participate in shouldering one's nation's proportionate share of the costs of a remedy for a problem to which one's compatriots past and present have as individuals contributed even unknowingly and faultlessly. The wrongful imposition arguments will now show why what I am labeling national policies of negative sovereign externalization are, in addition to being irrational and destructive from the perspective of humanity as a whole (and in the case of climate, self-defeating to a considerable degree at the national level), also a moral outrage well beyond the unfairness already shown by the earlier argument.

So far, I have mostly spoken loosely of the "costs" created by climate change, but there are costs and costs. That is, the carbon energy regime (1) generates economic costs that arise from damage that may not constitute wrongful harm and (2) generates fundamental wrongful harms, which of course also have their own attendant economic costs (assaulting someone is wrong in that it violates her right to physical security, but it may also generate medical costs for her, her medical insurance company, and her national health system).

Burning coal, for example, produces air pollution containing particulate matter, toxic chemicals that attack lungs and blacken historical buildings, and "acid rain" that blights forests, as well as the deaths of miners. Coal burning creates, among other damage, the need to wash and repair the surfaces of the buildings and install "scrubbers" in smokestacks to protect forests. The cost of restoring historic buildings is one large cost of coal combustion.[27] But the toxic chemicals in the air pollution that comes from burning coal also violate human rights by causing people to develop various lung diseases, including cancers, many of which are fatal. Causing someone to develop a fatal lung disease, especially so as to increase the profitability of one's business by sloughing off the costs of controlling one's own pollution, is inflicting a wrongful harm. Such lung diseases then add massively to the national economic costs of health care, but they most importantly violate the basic right to physical security against the infliction of fatal physical harms.

So coal burning generates both ordinary economic costs that may not constitute wrongful harms and clear wrongful harms like violations of the right to bodily integrity, which are accompanied by their own further economic costs. For shorthand, I will refer respectively to these two kinds of costs simply as "ordinary costs" and "wrongful harms." Much more could be said about the difference, but the relevant distinction here is that the imposition of ordinary costs is not in itself wrong in the same way that the infliction of rights-violating harms clearly is. This difference between ordinary costs and wrongful harms underlies a distinction in the moral significance of two analytically separable elements of negative sovereign externalization, which dumps both costs and harms from the creation of national wealth upon the rest of the world.

Observing this distinction requires wrongful imposition arguments, in turn, to branch into two versions. A. *Wrongful harms: violating human rights.* The main players in the carbon energy regime are contributing to wrongful harms by violating basic human rights.[28] As explained above, the burning of coal, cloaked as a necessity, has for centuries been violating the right to physical security, or bodily integrity, by polluting the air that people must breathe with carcinogenic and other toxic and otherwise damaging substances. And the combustion of all fossil fuels—coal, oil, and gas—is now violating the right to physical security by forcing climate change that brings deadly floods, deadly wildfires, deadly storm surges, deadly heat waves, migration of deadly disease vectors like mosquitos and ticks, and other fatal manifestations of a progressively disrupted climate. The chain of causation from combustion to climate instability is complex, although increasingly well understood.

The moral situation here, however, is entirely straightforward.[29] When one is violating rights, especially basic rights, one's duty is to stop entirely as soon as possible, bearing oneself whatever costs are involved. A commitment to rights entails a commitment to human solidarity to at least this extent, and the only good excuse for not desisting from the rights violations for which one is responsible is genuine necessity. When there were no affordable alternatives to fossil fuels, there may have been a period in which an excuse of necessity applied to the infliction of the harms they caused. But those days are long gone, and the combustion of fossil fuels has now become in fact an "avoidable necessity" because alternative energy sources are readily affordable and increasingly accessible.[30] Any accompanying benefits that accrue to those whose rights are violated do nothing to mitigate the wrong. Collateral benefits do not balance violations of rights.

B. *Ordinary costs: taking unfair advantage.* Are such direct human rights violations the only costs that are of moral concern? When the costs externalized, or dumped, are ordinary costs, and not inherently wrongful harms, is inflicting them on the remainder of humanity acceptable? No, the sovereign externalization of ordinary costs is also morally dubious in that it imposes costs on people of other nations arbitrarily: costs to which they have not consented and against which they are defenseless (and of which they are in many cases unaware). It is thus contemptuous of their dignity and disrespectful of their autonomy. Even if these costs were accompanied by collateral benefits great enough to create net benefit, it would still be unacceptable simply to force a particular package of costs and benefits upon people who were unable to decide for themselves whether to agree to it (in some cases because they were unaware of its terms). Even a net beneficial package may be incompatible with their own projects, plans, and way of life. Worse, when the externalization of costs imposes a net cost on another nation that was already worse off than the sovereign state engaging in the externalizing, the imposition of costs is especially objectionable because it coercively worsens international inequality.

Inevitably, to some extent most of humanity has benefited to some degree from the industrial societies in the "developed" nations—I am not assuming that the benefits of the industrialization process have been entirely retained inside industrial societies. But there certainly has not been a general distribution of benefits, while there has been a complete global distribution of the environmental costs and harms by way of the dynamics of the climate system. Once GHGs enter the atmosphere, oceans, soils, and vegetation, they circulate generally around the planet, and the emissions that enter the atmosphere produce effects that are worldwide, although not uniform. The lack of uniformity, however, bears no relation to the original source of the emissions; for example, temperatures are rising most rapidly in the Arctic and Antarctic, but the emissions causing this "polar enhancement" of temperature do not originate at the poles.

In fact, a common complaint made about climate change by ordinary people across the world is that benefits from fossil fuels and costs of climate change are misaligned with each other.[31] Who suffers the costs of climate change, for example, the costs of trying to adapt to sea-level rise, bears no relation at all to who receives the benefits from the emissions causing the sea-level rise. One might initially think that this is simply because as a matter of physical fact the climate changes do not occur where the emissions are released, but instead, for example, emissions from the temperate zones produce melting in the Arctic and the Antarctic.

However, while the location of the physical changes themselves is determined by the complex dynamics of the climate system, the distribution of the costs for dealing with this physical damage is the result of the workings of a social institution. This distribution of the costs, as distinguished from the physical damage, of climate change is mainly the result of a parochial and inward-looking human institution, namely the sovereign state, that ignores global physical dynamics and politically allows emitters shielded behind national boundaries to claim accountability for, and thus ownership of, the benefits of emissions while washing their hands of accountability for and ownership of the damage done. The distribution of the benefits and the distribution of the costs are clearly disproportionate. Can someone provide a compelling rationale for why most costs of damage caused by emissions ought to be publicly shared while most benefits produced by emissions may be privately retained? This is, on a global level, exactly the kind of externalization that economists regularly observe is irrational at the system level domestically and creates perverse incentives in domestic economies when practiced by firms—this is arbitrary externalization by sovereign states.[32]

Most importantly, it is deeply unfair. Why should humanity generally suffer the ill-effects of activities that they did not conduct, did not consent to have conducted, and may not benefit from to an extent commensurate with the costs they suffer? The unfairness is especially severe when the damage from climate change is inflicted on defenseless people. I have emphasized that "every state is a 'failed state' as far as climate is concerned."[33] Even sovereign states cannot effectively defend their citizens against the effects of emissions from other sovereign states—victimized states can only try to adapt to what they suffer at the hands of other states. For now, sovereignty protects the sources of the damage but not its vulnerable victims.[34]

The fundamental argument, then, is not at all that the historically greatest emitters have somehow emitted too much by emitting more than some proper share judged by some contentious standard, such as equal per capita emissions. Instead, the fundamental argument here is that emitters claim ownership of many benefits of their emissions while arbitrarily renouncing accountability for many costs of the very same activities.[35] Such accounting is misleading at a deep level about a state's own identity and its actual historical role and constructs a skewed and self-servingly positive interpretation of what it has done by acknowledging only the good and omitting all the bad in its understanding of itself. This is one-sided accountability: own the good and disown the bad. This is a distortedly partial picture of how one's nation has been behaving in the world. More importantly, this is disrespectful to everyone outside the national borders of any significant emitter and unfair to those who suffer a net cost, worst of all if they were worse off already before suffering the imposed additional costs.

The externalization of costs by powerful states is especially objectionable in the cases when they worsen the extent of international inequality in a morally unacceptable way rather than providing a net benefit. The standard I appeal to here is a relatively undemanding one: the avoidance of extremes of inequality consisting of benefits for some and net costs for others, imposed by human social practices like the systematic externalization of costs by sovereign states. In exactly which conditions is the infliction of more extreme inequality unacceptable?[36]

On the one hand, if a country comes up with some process that makes itself better off, but either does nothing to improve the position of any other country, or improves the position of one or more other countries somewhat but not as much as it improves its own position, the extent of international inequality increases. But there may be nothing wrong, other things being equal, with sometimes making yourself, and only yourself, better off, even though this will indeed increase the degree to which you are better off than those who were already worse off than you. The respective absolute levels of well-being are relevant to the limits on when this is acceptable, but these are complexities we must leave aside here.[37]

On the other hand, if a state improves its own position by a process that not only does not provide any net benefit to others but instead inflicts net costs on them without their consent, making them absolutely worse off than they were, as the state itself becomes absolutely better off than it was, and the others are helpless to shield themselves against its infliction of these costs upon them, making the infliction purely coercive, the state has *unilaterally* and *coercively increased* the extent of *inequality* between itself and them. This is patently objectionable and deeply disrespectful.

In cases in which the externalizer of costs makes other nations absolutely worse off without their consent and perhaps even without their knowledge, in order to make itself absolutely better off, the state has at worst treated that other part of the world entirely as a means to its own ends and at best shown those outsiders no respect. This is a paradigm case of a stronger party silently exploiting the vulnerability of a weaker party in order to pursue its own advantage. In Thucydides's immortal description, "The strong do what they can and the weak suffer what they must."[38] That such "power-plays" are currently standard practice in international affairs does nothing to make them less unsavory. It is a willingness to protect the vulnerable instead of exploiting or ignoring them that makes a society—including an international society—civilized. Exploiting or ignoring vulnerable outsiders displays contempt for human solidarity.

At best, the coercive and unilateral imposition of greater disadvantages by means of worsening the absolute position of others shows a complete lack of respect for them and their own ends. Viewed as a systemic practice with distributive consequences, such structurally imposed inequality produces a result with exactly the same shape that would come from redistribution directly from the worse off to the better off. In the case of carbon emissions that undermine the climate for others, the net reduction in the others' well-being is the result of the creation of new costs for them as they face the destabilized climate rather than the literal removal of some asset from them for redistribution upward. But they still become absolutely and relatively worse off if the source of the emissions retains the bulk of the benefits and sheds the bulk of the costs of the enterprises generating the emissions.

When the Chinese government pressures a poorer country to accept a Chinese-built coal-burning power plant as part of the Belt and Road Initiative, when that country could have been provided with a renewable energy source instead, because this benefits the incumbent Chinese government in the short term by providing employment for Chinese workers in politically influential but technologically backward Chinese state firms that are only capable of building coal-burning plants, the Chinese government maintains political stability for itself by inflicting upon the rest of the world severely climate-worsening infrastructure that can be expected to undermine the climate with carbon emissions for forty to fifty years—until around 2060 or 2070 if put into service in 2020.[39] The same is of course true, to a lesser degree, of the US federal government's indulgence of utilities that refuse to retire gas-burning generation: for (relatively small and short-term) benefits for domestic constituents (specifically, for the firms with vested interests in obsolete gas technology and for the politicians who receive their campaign contributions), it imposes (large and long-term) costs on everyone else (and long-term on itself as well, since the policy is self-defeatingly irrational).[40] These are contemporary continuations of the historical pattern of coercive and unilateral imposition of worse international inequalities.

When the US Department of State declared in 2018 that it "will not allow climate agreements to be used as a vehicle to redistribute wealth," what it actually did was very different from what it said it was doing. It was refusing to acknowledge our responsibility as Americans for the contribution made by the emissions from our own industrialization to the climate changes throughout the planet. Myopic statements from the US federal government over the last quarter century during international negotiations concerning climate change have adamantly insisted that all wealth currently controlled by the United States is beyond all doubt rightfully held and that any difficulties with the changing climate faced by anyone else anywhere in the world, including all the poorer nations who have still not industrialized to this day, are entirely their own problem, with any request for assistance being an underhanded attempt to "redistribute" what rightfully belongs to Americans. This is a preposterously simplistic and distorting position that erases the complexities of the real history and reflects no inkling of solidarity with—or even respect for—the rest of the human race.[41]

On the one hand, if it is wrong for the powerful to take advantage of the powerless by imposing greater inequalities, then countries like the United States ought to correct the historic wrong by reversing this inequality at least to the extent to which they imposed it through externalization of the costs of their own industrialization on others who were already worse off. The most relevant way of implementing this correction is by contributing to the costs of bringing climate change under control, that is, taking on the very "burdens or financial pledges in support of the Paris Agreement" that the US federal government arrogantly renounced.

On the other hand, even if there somehow were nothing wrong with the powerful having taken advantage of the powerless decade after decade by releasing national emissions into the shared air, land, and sea—if the wrongful imposition argument were misguided—the case would simply revert to pure fairness, the first of this chapter's three arguments. The costs of bringing under control the climate change caused by those emissions must be paid, so someone must bear these costs. On whom could it possibly be fairer for those costs to fall than on those who contributed to causing the problem, held on to the bulk of the benefits of doing so, and allowed much of the costs to fall upon others who benefited less? As I noted twenty years ago, "If there were an inequality between two groups of people such that members of the first group could create problems and then expect members of the second group to deal with the problems, that inequality would be incompatible with equal respect and equal dignity. For the members of the second group would in fact be functioning as servants for the first group."[42] To insist on retaining benefits while refusing to bear the burdens created by the same process can only be unalloyed greed devoid of respect for anyone who is not a compatriot.

Conclusion

Other things being equal, it seems clearly fairer for those who have contributed most to the creation of a problem to bear much more of the burden of dealing with the problem than those who have contributed least. Moreover, other things being equal, it is evidently fairer for those who have benefited most from the creation of a problem to bear much more of the burden of dealing with it than those who have benefited least. While the present descendants of those who contributed most to the creation of the polluting carbon regime did not themselves contribute to its creation, we are contributing to its perpetuation in the present and to the worsening of climate change in the future, insofar as we simply use the inherited energy regime and do little to transition beyond it; and we enjoy the benefits (and the ability to pay[43]) inherited by present members of the nations whose earlier members did create the problem.

It may seem unfair that any generation should have to bear the burden of fundamentally restructuring the global energy regime. When James Watt designed the steam engine in 1784—the event from which Paul Crutzen dated our entry into the Anthropocene[44]—burning coal seemed simply to be an ingenious way of generating copious amounts of steam and thereby unprecedented amounts of energy. Who knew that coal would turn out to be the most climate-disruptive source of energy available? But it turns out that the CO_2 from fossil fuels that reaches the atmosphere holds in unwanted heat that used to escape from the planet, so we have no choice but to stop injecting the CO_2. This requires an expensive transition between energy regimes, and someone will have to pay.

The fundamental contention here is that what has happened between 1784 and today is very nearly the most unfair process imaginable and cries out for a robust correction. One portion of humanity—the "Developed States"—has reaped the vast majority of the benefits from the invention of the steam engine and the Industrial Revolution generally while allowing the costs, including rights-violating harms and increasing inequalities, to be spread globally. Most individuals will suffer from climate change, although not uniformly and not in any proportion to their contribution to causing it. It is perhaps unfair that the benefits have been narrowly held while the costs have been widely disbursed. It is certainly deeply unfair that the benefits have been narrowly held by those who have inflicted the damage on everyone, while the costs, including severe harms—such as loss of life, health, or home—descend randomly upon all. The heart of what is objectionable is, if you like, a specific conjunction of benefit from the problem, contribution to the problem, and infliction of harm: those who are the source of the dangers from the disrupted climate suffer least from those dangers and keep more of the benefits from the activities that are disrupting the climate by failing to shoulder the costs of what must be done to head off the far worse dangers inherent in the persistence of the now-dominant carbon energy regime.

It would be difficult to concoct a more strikingly unfair arrangement than the energy-business-as-usual. Contribution to solving the problem ought to bear some relation to contribution to creating the problem, especially when those who have in fact created the problem have benefited so handsomely from doing so and those who suffer most have made little or no contribution to the problem. It is unfortunate that anyone must pay, but granted that someone must, the best that we can do is to assign the costs to those whom it is fairest to charge from among those who are in fact available to be charged. Some third-party payer—an imaginary philanthropic foundation from a distant planet, perhaps—would be preferable to either of the real choices. But it would be outrageously unfair if those of us who thus far in our lives may have captured, perhaps inadvertently, the benefits of our industrial society while continuing generally to disburse the harms should now knowingly insist on continuing to cling to the maximal benefits from massive carbon emissions while inflicting maximal harms on people generally by refusing to bear a substantial share of the costs of escaping the energy regime that is disrupting the climate while benefiting and enriching us. We would be twisting inadvertent unfairness into conscious exploitation. The Owl of Minerva has flown—we can see exactly what we are doing if we do not avert our eyes.

One government whose approach to climate change seems to express such conscious exploitation of the rest of the world has been the US federal one. US efforts so far on climate change at the federal level have always been pathetically weak—under the Trump Administration, they were hostile to humanity outside the United States and generally perverse. It is past time for the US national government to become fully serious about reducing carbon pollution, and initially we need to know only this: who ought to act now. We can later figure out when various parties have done enough and give them credit for any past performance in reducing their contribution to the threat once the worst dangers are averted. We do not need rankings among governments of their contribution to the problem; and arguments about who is worse than whom can at this point easily become a dangerous distraction, which entertains theorists but confuses the public. Many other governments—notably the Chinese, Russian, and Brazilian—also need to do vastly more than they are doing.

My argument is basically a consistency argument about fairness. The federal government of the United States, like many national governments, claims for its present and future citizens most of the fruits of the activities of its past citizens—it claims national ownership of, for instance, the benefits of the industrialization of the United States, including the vast infrastructure and capital left behind. In consistency, if most of the benefits of the past belong to the United States, so should the corresponding costs. But one very significant cost, the enormous damage done to the stability of the global climate system by carbon pollution, has in fact been socialized universally. The people of all nations and all generations are increasingly suffering the effects of the vast GHG emissions produced by the process of US industrialization and the continuing maintenance of US postindustrial society. Similar evasions of accountability by other governments, like the Chinese, Russian, and Brazilian, excuse no one, them or us.

3

Engagement across Distance and Engagement across Time

Any man's death diminishes me, because I am involved in mankind. And therefore never send to know for whom the bell tolls; it tolls for thee.[1]

The next few decades offer a brief window of opportunity to minimize large-scale and potentially catastrophic climate change that will extend longer than the entire history of human civilization thus far.[2]

We are scientists who have been studying the Amazon and all its wondrous assets for many decades. Today, we stand exactly in a moment of destiny: The tipping point is here, it is now.[3]

India's poor have rights that can be satisfied only through economic development[4]—only by, for one thing, making electricity available to the hundreds of millions of Indians whose "energy poverty" extends to the extreme of having no access at all to electricity. But massive burning of the world's dirtiest coal, which is the most plentiful source of energy that India has in the past had the infrastructure to exploit, makes the air in major Indian cities more polluted than even the notoriously polluted air of Beijing, and in addition greatly reduces the odds of limiting climate change across the globe to levels that will not be dangerous for billions of people across many generations. What should be done? How much of it, if any, is a problem for those of us who live in a wealthy country? Do we share in the responsibility to find a way out of India's dilemma? If so, why? And how much?

This chapter will show concretely in the case of climate change that both international justice and intergenerational justice rest upon human relationships that are far less distant than assumed by conventional thought and established practices. Causal webs tightly link persons across both space and time. Both the fossil-fuel energy regime that is driving climate change and the measures necessary to make the transition from that regime into an alternative energy regime impinge deeply upon the well-being of contemporaries who have chosen neither the regime nor the transition, linking them across space to those who control the regime and the fate of the transition. Similarly, the fates of persons in the distant future are in the hands of people living now because in the case of the climate system the date-of-last-opportunity to prevent disasters from becoming irreversible often occurs decades or even centuries earlier than the start of the disaster itself. This is powerfully illustrated by the evident irreversibility of the melting of the West Antarctic Ice Sheet, which will ultimately produce catastrophic rises in sea level across the globe, affecting hundreds of millions of people.

Some responsibilities of international and intergenerational justice are entirely forward-looking and completely independent of the historical responsibilities based on past contributions to climate change that we have just examined in chapter 2. The applicability and strength of other responsibilities to do international and intergenerational justice, by contrast, rest on those historical contributions to the problem. A main goal of this chapter is to lay out clearly where such connections do and do not hold.

The Phenomenology of Agency

In a probing 1995 exploration, Samuel Scheffler suggested that our inherited conception of responsibility rests in part on what he called "a complex phenomenology of agency": a set of assumptions about what matters about what we do and do not do.[5] Specifically, he saw that among the concepts deeply ingrained in our understanding of the world are the following two assumptions: the near effects of our acts are more important ethically than the remote effects, and the effects we produce as individuals are more important ethically than the effects we produce as members of a group.

The first assumption, I think, reflects its origins in somewhat less globalized times, although we should not exaggerate the simplicity of earlier times.[6] That the near effects of one's actions are more significant than the remote effects made sense in earlier centuries when in fact one's most powerful influence—for most people most of the time, one's only influence—was local. In those times, if one wanted to have some effect on people elsewhere than where one lived, one needed to go to that other place in order to act there. Now a person's greenhouse gas emissions over the course of her life in Ireland contribute to the sea-level rise on the coast of Africa, and a tweet from Peru can cause a riot in Moldova.

The second assumption, that the effects one produces acting alone are ethically more significant than the effects one produces as a member of a group, was, I suspect, never entirely true because, apart from utter hermits, individuals are normally embedded in communities—even homeless people and refugees affect others. The conventional assumption may be largely explained by individualistic ideological commitments. It is not true that the effects of what one does by oneself are always more important than the effects of one's participation in institutions and practices like being a lifelong customer in a consumer society based on the energy from fossil fuels, which is burned in the production of consumer items, the transportation of consumer items, and the use of consumer items.[7] Climate change is indeed a giant example of *remote*—indeed, planetary—effects that are also the *group* effects of individuals participating in an international energy regime that is dependent on coal, oil, and natural gas. And climate change is a powerful reason why we need to try to follow Scheffler's early lead in rethinking our understanding of responsibility, which I hope this book will extend.

Any normal person whose conception of herself extends beyond personal material comforts and psychological pleasures to include attachments to children and grandchildren and to practices and institutions—be they football or opera—that endure across time and are likely to be enjoyed by children and grandchildren too has broadly self-interested reasons for caring about climate change because of its potential to disrupt ordinary lives and the travel habits of both receivers and sopranos. Except for sociopaths who literally care only about themselves—and even they should worry that they may find their house flooded or their food more expensive—ordinary humans who care about at least some of the people and activities that survive them have good reason to take action to limit climate change in order to preserve what they value.[8] But consider instead the billions of other people, those individually unknown to us—in many cases, spatially distant—for whom we have no reason to feel affection and whose activities and practices we may have no particular reason to care about—the vast majority of other people in the world. Do we in the wealthy countries have any responsibility toward them with regard to climate change? If so, why? And how much?

The Danger of Transition

The Russian federal government seems devoted mainly to increasing exports of gas and oil,[9] and my own US federal government is for now largely immobilized by the votes of senators and congresspersons who appear to have been bought for blocking actions against climate initiatives by campaign contributions from fossil-fuel interests,[10] but most societies are now to some degree engaged in some kind of effort to respond to climate change. So we need to reflect on both the climate problem itself and the proposed responses, because of course every response to a problem comes with its own problems.[11] We should consider any responsibilities we might have to distant strangers regarding either climate change itself or our responses to climate change. Recall two of the notorious general features of climate dangers.

First, the distribution of the dangers bears no relation to the distribution of the benefits from the emissions that are causing the dangers. Sea-level rise, for example, will affect those who live near sea level—thus, hundreds of millions in Bangladesh and India who do not even have electricity will suffer from encroachment, flooding, and storm surges partly caused by electricity generation from coal.

Second, the death tolls from storms are heavily affected by the wealth of the society that suffers the storm: hurricanes that hit the United States kill relatively few people—sometimes no one—because buildings are sturdy, emergency broadcasts are effective, medical care is excellent, and so on—while cyclones that hit the Philippines regularly have high death tolls.[12] Anyone may be affected by climate change, but, generally speaking, the poor will suffer the most.

The dangers that will come from our response to the climate dangers may be less familiar, so it is worth noting them slightly less briefly. By far the most important factor in causing climate change is the accumulation in the planet's atmosphere of CO_2 from the burning of fossil fuels. And it turns out that because CO_2 remains in the atmosphere for almost unimaginably long times—between roughly 10% and 25% of it stays for several hundred thousand years[13]—the crucial factor is the *cumulative* amount of CO_2 emitted since around 1750, the vast bulk of which still remains in the atmosphere. The atmospheric physicists have established—and there is now very wide consensus on this—that there is a fairly specific *cumulative* carbon budget for any particular probability of not exceeding the pre–Industrial Revolution temperature by any given number of degrees.[14] An up-to-the-second calculation of total cumulative CO_2 is available.[15]

So if we are serious about even the unambitious goal of not exceeding a 2°C rise, it is critical for global emissions to peak and then start down sharply very soon, which is physically and technologically possible. It simply has to be made politically possible by acts of determination and will that unseat incumbent politicians blocking action. A sharp decline in emissions would push the date for exceeding the cumulative budget back further into the future. By about 2050, CO_2 emissions from energy use must completely stop, which means humanity must exit the fossil fuel, or carbon, energy regime entirely and use nothing but alternative energy, that is, noncarbon energy—anything but fossil fuel.[16] This is because if we keep adding to the cumulative total, the atmospheric concentration must exceed the carbon budget for 2°C.

This should all play out, then, well within the lifetime of the current under-forty-year-olds. The next decades will be an amazing time to live on this planet, and no one should be bored. Today's young will witness one of history's greatest struggles that will result in either one of humanity's most glorious triumphs—a successful Energy Revolution eliminating fossil fuels and their carbon emissions—or one of humanity's most dismal failures, the triumph of the entrenched carbon energy regime, the leaders of which are resisting the needed transition with stealth, trickery, and lies,[17] and the consequent coming of yet more dangerous climate change.

What I want to highlight here are the dangers that come, not with the climate change itself, but with our response to it. Fossil fuels, especially coal, which is by far the worst for both carbon emissions and lethal surface-level air pollution, need to be made extremely unattractive extremely fast—80% of the known reserves of coal must be left in the ground to avoid overshooting the goal of a rise in temperature of no more than 2°C.[18] Governments could simply regulate further extraction of coal, oil, and gas and eventually ban it—supply-side measures—but no democratic government has so far had the guts to regulate fossil-fuel use to save the climate, as was done through rationing, for instance, to save countries during World War II.[19] So most of the proposals for reducing fossil-fuel use involve political action to employ market mechanisms to make it more expensive until consumers themselves choose to turn away from it to other energy sources—demand-side measures. It can be made more expensive in either of at least two ways: a carbon tax or so-called cap-and-dividend. I have nothing to add here to the disputes among regulation of extraction, cap-and-dividend, and carbon tax, except the obvious point that political leaders need to face up to one or more immediately.

If politicians can be pressured by concerned citizens into acting, fossil fuels will start to become more expensive, which is good as a means of reducing combustion. But this is where the danger for developing countries from the necessary response to the dangers of climate change begins. And this danger means that the response to climate change must not have only one dimension. The danger is disrupting development by depriving the poorest of the only energy source that they could until recently afford: fossil fuel. The low prices for fossil fuels in the past have been partly the result of two mistaken political policies.

First, many poor country governments to this day heavily subsidize fossil fuels as part of a grossly inefficient tactic to promote development, but the dangers of climate change make this a deeply misguided policy that needs to end immediately, even if it had been an efficient mechanism before we understood climate change. The world total of fossil-fuel subsidies in 2015 was $4.7 trillion.[20] Amazingly, G-20 governments liberally subsidize fossil fuels, with both producer subsidies and consumer subsidies, in spite of their rhetoric in support of capping temperature rise.[21]

Second, the price of fossil fuel is as artificially low as it is also partly because extreme deficiency in political regulation has permitted most of its enormous health and environmental costs to be externalized, that is, not covered by the price for the purchasers of the energy but dumped upon society generally—we all pay for the hospital expenses of fellow citizens with lung and heart diseases from breathing the pollution from the burning of coal—and dumped on beyond fellow citizens to humanity generally through the "sovereign externalization" that we considered in chapter 2. Neither a carbon tax nor cap-and-dividend might be necessary if the price of fossil fuel covered the huge health costs of the lung diseases and shortened life-spans caused by the air pollution from coal burning and its monumental environmental costs, above all climate change itself, but also the despoliation of land and pollution of water from the extraction of all fossil fuels, including coal mining, especially strip mining, and fracking for both gas and oil, which both wastes and pollutes shared water supplies.[22]

Much of the environmental damage is hidden from the view of us in affluent countries because it is in the oceans or in the developing world, such as the gigantic mess Shell has left in Nigeria; the Niger Delta has been called "the most oil-ravaged place on the planet."[23] Shell has failed for decades to clean up massive oil leaks, leading to the deaths in protests by the Nigerian people against Shell's inaction, which inspired Nnimmo Bassey's poem:

> We thought it was oil
> But it was blood
> We thought it was oil
> But it was blood.
>
> They may kill all
> But the blood will speak.[24]

In general, politicians have given fossil-fuel corporations by far the biggest free ride from respect for the environment of any firms in human history—plus tax breaks! Is it possible that many politicians are bought and paid for?[25]

In any case, if governments belatedly act against climate change by pricing carbon through either carbon taxes or cap-and-dividend, the prices of fossil fuels will finally rise, and what has in the past been the most affordable energy for the poorest will become more expensive, will specifically become unaffordable for many of the poor who can afford it now, and will rise even farther out of reach for those who already cannot afford it. We have no choice but to limit climate change, but if governments do it by raising the market price and do nothing else, the implications for development are ominous. Plainly, then, we must not choose to price the poorest out of the energy market and simply leave them as helpless as we would have then made them. Efforts to limit climate change cannot be pursued in isolation. We must simultaneously pursue complementary policies that avoid undermining development and forcing those who are already humanity's worst off to become even worse off. Which policies?

Contemporary "Strangers" and the Affordability of Alternative Energy

Fairly evidently, it is our responsibility to protect the poorest from the results of our own choice of policy instruments—for example, a politically driven rise in fossil-fuel prices—by supporting the rapid spread of alternative forms of energy by the most effective means available. This could mean, for example, feed-in tariffs, as used early on by Germany, or special tax concessions for entre-preneurs and/or early adopters of new technology, or increased research and development funds for universities, or subsidies for community electricity microgrids supplied by solar panels, or other techniques that a theorist like me should be the last person to advise about.[26] The crucial constraint is that the alternative forms of energy must be made both accessible and affordable to the poorest in the developing world soon to replace any electric-ity that could otherwise be generated by burning coal, oil, or gas.

Thus, a primary focus of the developed world should be strate-gies designed to put the alternative energy into the hands of the poor, not only the acquisition of alternative energy ourselves, so that investment in the developing world is not wasted on infrastructure for fossil fuels, like coal-burning or gas-burning electricity-generating plants, that would lock in fossil-fuel use for additional years and would in any case have to be replaced sooner rather than later. With prompt assistance, poorer countries can leapfrog over the carbon infrastructure that wealthier countries are already stuck with. Replacing our own polluting infrastructure will be more expensive and will take longer, although we should certainly avoid adding any more, such as the many proposed new pipelines for either oil or gas that are misguided diversions of capital investment. Gas is not, contrary to corporate greenwash, a "bridge away" from fossil fuel; it is a fossil fuel.[27] If the public subsidies come off all the fossil fuels, including gas and its pipelines, and are instead put on alternative energy for a transition period, there is no reason why noncarbon energy that does not undermine the climate cannot continue to become increasingly competitive in price with damaging and now grossly underpriced fossil fuels. Indeed, the prices of, for example, solar and wind have plummeted in recent years and are already in most places competitive in price with fossil fuels even with the latter's misguided public subsidies.[28] Analyses by the rigorous Carbon Tracker Initiative show that new investments in renewables are now cheaper than new investments in coal in all major markets (even Australia, with all its coal), and over half of coal plants operating today cost more to run than building new renewables now does.[29]

Not only will sufficiently affordable alternative energy help to overcome "energy poverty" in the developing world and head off increases in emissions there, but it can directly assist in eliminating current carbon emissions by making replacements for fossil fuel readily available in the developed world. In the affluent world, much of our energy is literally wasted or consumed by unnecessary services and products, and by simply reducing waste and indulgence, we can fairly easily use less carbon energy without replacing it with anything.[30] But where people, including those who do not appreciate the seriousness of the dangers from climate changes, are unwilling or unable to reduce energy usage, they can have a price-competitive replacement for fossil fuels.

In sum, in order to stay within the cumulative carbon budget for the planet as a whole, carbon emissions need both to come down sharply in the developed world but also not go up very much in the developing world, while development nevertheless accelerates sustainably. At both ends of the line, accessible and affordable noncarbon energy is crucial. For once at least, the same response, alternative energy, can serve (some of) the desires of the rich and all the needs of the poor. But smart targeted initiatives are needed to make it accessible to the poor quickly enough—prompt affordability for the poor is not a challenge that unaided market mechanisms can possibly meet.[31] I have tried, then, to be reasonably concrete (for a theorist) about what needs to be done, although I have not tried to say which specific forms of alternative energy should be encouraged or exactly how, since these do not need to be top-down decisions.[32]

However, it may be useful to illustrate a concrete possibility. India is in fact proposing a highly ambitious initiative, "One Sun One World One Grid," aimed at ultimately creating a single planetary electricity grid, using subsea High-Voltage Direct Current (HVDC) cabling for the connections between countries. Grid connectivity would leverage the time differences between countries to manage the intermittency of low-cost solar- and wind-power generation: "a 1,000 km subsea HVDC connection from Oman to Gujarat would take ultra-low cost solar electricity . . . in the middle of the Oman and UAE afternoon and deliver it into the evening peak in Dhaka."[33] During Mumbai's night, the sun may be shining in Bhutan and the wind blowing in Nepal. Such a grid would obviously require extensive international cooperation, and at its maximum, it would literally and physically connect much of humanity across national boundaries. As the Institute for Energy Economics and Financial Analysis in Cleveland observes, "Virtually unlimited low-cost renewables are available today at massive scale in India, with zero inflation indexation for 25 years. This is entirely sustainable power producing no carbon emissions and no air or water pollution. Deflationary green exports can realistically entirely replace expensive, inflationary, polluting fossil-fuel imports, with vision and international co-operation."[34]

We need to move toward a systematic look at the various intersecting responsibilities. Why should rich countries cooperate with such projects (not necessarily precisely this one)? Why do we in the rich countries share in the responsibility for the dissemination of alternative energy within and among developing countries? How did this become a task that partly falls to us? I will brush over familiar reasons that have nothing to do with climate change. For example, if every human being is to have a meaningful basic right to subsistence, then all circumstances require an international division of moral labor sufficient to see that those who are now unable to provide for their own subsistence are enabled to become capable of providing for themselves, which will involve in many cases the external provision of resources not now available locally and therefore dependent on the rest of us.[35] But some less familiar, but important and interesting, reasons specifically concern climate change.

The Indian government is entitled to give high priority to reducing poverty in India. India, which arguably is of all the countries in the world the most threatened by climate change, will cut its own throat, however, if it injects large additional amounts of CO_2 into the atmosphere and contributes to humanity's exceeding the cumulative carbon budget for some "nondisastrous" amount of temperature rise, not to mention creating dangerous amounts of ground-level coal pollution that will continue to multiply cases of lung and heart disease within Indian cities.

It is vital, however, to remember that additional Indian carbon emissions will be such a dangerous problem for the planet only because of past and present emissions from developed nations and more recently from the continuing and sharply worsening surge of overwhelmingly coal-driven development by China.[36] The cumulative greenhouse gas emissions of the United States,

Great Britain, and China each considerably exceed the cumulative emissions of India at this point.[37] India's contribution to the fact that humanity as a whole may soon exceed the global cumulative carbon budget has so far been relatively minor—far less than the contribution of the United States or the European Union, and per capita, only a fraction. But like China a couple of decades ago, India is on the threshold of rapid, massive increases in energy consumption. What energy sources will it choose?

Now, as Indian and Chinese leaders argue—and indeed I have argued for twenty-five years—it would be wildly unjust for us in the rich countries to say, "Sorry, but our own development and our own continuing enjoyment of affluence have used up most of the cumulative carbon budget so there is no room within the budget for the emissions that your escape from poverty would generate, so you will unfortunately simply have to remain poor."[38] (And no one would listen to us, in any case.) On the other hand, the cumulative carbon budget for a particular temperature rise is not an artifact that can be modified by choice. The cumulative carbon budgets for various degrees of climate change are unrelenting features of the dynamics of the climate of this planet. If we exceed the cumulative carbon budget for a reasonable probability of a given temperature rise, the actual temperature rise will probably be larger, and the concomitant climate change will be more dangerous. There is no overdraft privilege for the planetary carbon budgets, however much we might wish there were.[39] It is conceivable that the scientists have miscalculated them—specifically, overestimated our allowance[40]—but the politicians cannot fudge them.[41]

It would be profoundly unjust if India—and Africa, much of whose development is still to come—could not develop because the carbon emissions budget is about to be used up by the already affluent. Nevertheless, development produced at the price of exceeding the emissions budget would simply not be sustainable— it would in fact undermine itself. This was the global dilemma of the twenty-first century before the prices of renewable energy fell so rapidly and imaginative solutions, including massive batteries and innovative electricity grids using AI, were found for the intermittency in any one place of wind and sun. Therefore, development ought to proceed, but proceed sustainably, which can only mean with sharply declining carbon emissions, not rising carbon emissions, as in the planet-destructive case of China's persistence in building coal-burning, electricity-generating plants both in China and abroad (through the Belt and Road Initiative).[42] The only good alternative is a rapid scaling up of noncarbon energy within, and perhaps among, developing countries—not necessarily as ambitious as "One Sun One World One Grid"[43]—plus measures to liberate women, like education, jobs, and access to affordable contraception, that will slow population growth.[44] These are daunting challenges politically and technologically, but responsibilities to see that something like this happens are shared by the developed countries, then, for at least three conceptually separable reasons.

The first of these reasons is grounded in causal responsibility for climate change itself. Thus, it is grounded squarely in the national historical responsibility analyzed in chapter 2. However, nations also have responsibilities of international justice with bases other than the damage done by their own past behavior. The second reason depends on forward-looking negative responsibility to avoid, in our effort to deal with climate change, causing damage by a policy of making fossil fuels more expensive and scarcer within our dominant energy regime without providing adequate alternatives to fossil fuel. The third rests on another closely related negative responsibility not to exploit the vulnerability of the poor and weak, which is generally independent of historical responsibility. This duty is especially stringent, however, to the extent that one has contributed to creating the vulnerability exploited, which depends on the extent of one's historical responsibility. To create vulnerability and then exploit it is "compound injustice."[45] We look at each of these three reasons in turn.

First, we have seen that India's potentially looming emissions would only be so dangerous because our emissions preceded theirs. The Indian—or later, African—trunk well might break the camel's back, but only because the camel is already carrying the American, European, and Chinese trunks, as we saw in chapter 2. As also explained there, it is essential to distinguish the purely physical from the economic determinants. Physically, on the one hand, additional emissions anywhere on the planet raise the general probability of worse weather in various otherwise physically distant places—the retention by greenhouse gases of more heat energy on the planet than in the previous millennia of human existence drives volatile weather. The Indian monsoon, for example, has recently been departing from its previous patterns in ways that are desperately threatening to the supply of drinking water in India.[46]

Economically and politically, on the other hand, it is our choice whether we continue to practice the sovereign externalization analyzed in the wrongful imposition arguments in chapter 2, as well as the simple unfairness analyzed in the pure fairness argument, by asserting that all costs for mitigation, adaptation, and loss resulting from climate change in India must be borne by India. We could persist in the "fateful and contentious assumption" that "whatever problems arise within one nation's territory are *its own*, in some sense that entails that it can and ought to deal with them on its own, with (only) its own resources."[47] This is, however, an outrageous assumption given what we know about our own centuries-long causal contribution to the buildup of CO_2 underlying those problems—this would be handwashing squarely in the ignoble tradition of Pontius Pilate.[48]

Second, it could be our decision to discourage carbon emissions by taking measures to raise the price of carbon that would risk pricing poor Indians out of the market for what might remain for some of them their only affordable source of energy, if we simultaneously allow initiatives like "One Sun One World One Grid" to fail for lack of international support. As alternative energy continues to gain a greater price advantage and to become increasingly plentiful, measures that raise fossil-fuel prices should negatively affect progressively fewer people.[49] Support for alternative energy can, then, reduce the damage done during the transition between energy regimes.

Meanwhile, both these actions of ours—our large contribution to the problem and our choice of a market-based solution, if that is what we choose—could deepen India's plight and make our actions partly causally responsible for the resultant harm. A principal reason why financial support for India's transition to alternative energy is not a "foreign aid program" is, then, that at bottom it constitutes taking action in order to refrain from severely harming, not positively helping—from additional severe harming, if one accepts the wrongful imposition arguments. Replacing carbon-based energy with alternative energy promptly—even preemptively—can avoid the harm of energy deprivation during the unavoidable energy transition. This support is necessary in order to create an escape route for them from the corner we will have forced them into by our emissions, including the many unnecessary ones that continue unabated today.

This second reason we bear moral responsibility rests, in turn, on a rock-bottom principle: avoid depriving people of necessities without which they are helpless (however good your ultimate goal—even limiting climate change).[50] The necessity in this case is adequate energy that it is not self-defeating to use. Once we were unknowingly contributing to the deprivation produced by climate change. Next, we could be contributing to the deprivation produced by solutions to climate change. We ought to provide protection against further future deprivation as well as grant support in tackling the present deprivation that we already helped to create.

Thirdly, and very closely related, if we allow the world's poorest to bear the brunt of the transition out of carbon energy and into alternative energy, we are, in the common phrase, balancing the energy transition on the backs of the poor. We would in effect be producing an upward redistribution of wealth from the poor to the rich by making those who are already among the worst-off members of humanity suffer the costs of the global transition between unsafe and safe energy regimes rather than allocating some of our own wealth to invest in lightening their burdens, thus keeping all the benefits of the old energy regime for ourselves while dumping the pain of the transition to a new energy regime on them.

This is compound injustice because a costly energy transition is required in order to escape the global energy regime that we early industrializing nations created. Leaving the poorer nations to pay for their own transition out of our carbon energy regime would clearly be compound injustice because it would make the worst-off worse off still than they are, as well as far worse off than we are, as a result of their being left to wrestle on their own with a solution for a climate problem they did not create, but we did.

Raising the price and reducing the supply of fossil fuel would likely reduce carbon emissions in India, which in itself is a good thing. But in isolation, these measures could deprive some poor Indians—and Africans, and others—of any source of energy they could afford, and thereby effectively deprive them of energy. We would be reducing our damage to the climate by trapping the world's poorest in misery that they lack the means to escape.[51] As the agency Practical Action observes, "A life without access to energy is a life of drudgery."[52] This upward redistribution resulting from responses to climate change relying on the demand-side market mechanism of higher prices for fossil fuels or the supply-side mechanism of restrictions on the extraction of fossil fuels would be in addition to the upward redistribution already resulting from our contribution to the creation of the problem of climate change itself, which will force India to incur enormous expenses in dealing with, for example, sea-level rise and storm surges along its vast coasts and disruptions to the monsoon in its interior. The rich got energy and became richer; the poor will get sea-level rise and become poorer, if we treat them as no concern of ours and thereby deepen global inequality.

We have now seen reasons of justice why we have the responsibility to perform three specific tasks: we should not use so much more fossil fuel ourselves that our emissions do not leave enough of any of the remaining cumulative carbon budget to allow some of the poorest to rely unavoidably for a little longer on inexpensive fossil fuel, while they lack access to any other energy; we should see to it that in the course of preventing the cumulative carbon budget from being exceeded, the poorest are not priced entirely out of the energy market; and we ought not to orchestrate an unjust transition that creates benefits for the richest but costs for the poorest. Failure at either of the first two tasks deprives the poor of necessities; and failure at the third unjustly exploits their vulnerability and exacerbates inequality—it inflicts compound injustice. And we have seen that an alternative policy is readily available: we can contribute substantially to making alternative energy accessible and affordable for today's poorest through supporting initiatives that provide them, for instance, electricity from renewable energy (irrespective of whether that becomes part of an ambitious global grid). We must keep "the poor of the future in mind, but also today's poor, whose life on this earth is brief and who cannot keep on waiting."[53]

Future "Strangers" and Tipping Points

Pope Francis is correct that we should not deceive ourselves into thinking that we must choose between today's poor and tomorrow's poor. The poorest in the future will be even more vulnerable to us than the poorest living today, because our relationship with future generations is totally asymmetrical. We control the kind of world in which they will have to begin to shape their lives, but they have no way to influence our decisions about the world we leave them—indeed, even which individuals will be born may be determined by us. In this context, the people of the future, whoever the individuals turn out to be, need at least two complementary things from us.

First, they need entrenched sustainable development now so that the poorest of the future can have the resources for better lives than their ancestors, today's poorest. This is another powerful reason why we must urgently get alternative energy into the hands of today's poor rather than allowing them in desperation to use more fossil fuel and undercut their own long-term development by injecting more CO_2 into the atmosphere. The best way to improve the lives of tomorrow's poor may be to improve the lives of today's poor, who will bring tomorrow's poor into the world and provide them with what nutrition and education they will get—further reason not to think one should or could choose between the current poor and the future poor.

Second, the future poor need for us to bring climate change under control and to fix as firm a ceiling as we can on the disruptions that climate change will bring them, especially in order to prevent disasters that it will be too late for them to prevent and possibly too late for them to adapt to.[54] The necessity for us to act now if they are to be spared extreme climate change later is the basis for the extreme urgency of immediate action.

In chapter 1, we noted three reasons why those alive now can reasonably be expected to make an exceptionally robust effort on climate change: (1) future generations will very likely face burdens and dangers greater than ours, (2) the worsening dangers are currently unlimited, and (3) less effort by us may well allow climate change to pass critical tipping points. In sum, the burdens and dangers for future generations will probably be worse and are now worsening, are still without limit, and are potentially unbearable. We also saw that historical context matters in that, in general, some tasks are not deferrable to another time or transferable to another generation.[55] These tasks must be done now, or they can never be done, because the window of opportunity to do them will have closed. Obviously, responsibilities to respond to the third kind of consideration, the danger of passage beyond tipping points, are utterly nondeferrable. And the failure to set a limit entails a failure to prevent the passing of further tipping points. It may be worth probing these grounds of responsibility a bit more, even finally in some quite tentative ways.

The Date-of-Last-Opportunity to Prevent Disaster

The critical time to consider, then, is not the date at which a disastrous event will begin to occur, which may not be until some considerably later time in the future, decades or centuries ahead. The critical point is what we can call "the date-of-last-opportunity," that is, the last time at which it is—or, was, if the opportunity has already been missed—still possible to prevent the disastrous effect in question. The tipping point may lie decades or centuries prior to the emergence of the disaster because of the long time lags in the climate system between lock in and occurrence.

Some scientists have recently put the point in terms of the relation between what they call "intervention time," which is the amount of time left during which an intervention could still occur (or the amount of time left before an outcome becomes physically locked in), and "reaction time," which is the amount of time it would take humans to carry out the relevant intervention. Their point is that "if reaction time is longer than the intervention time left . . . we have lost control." They illustrate the point by assuming that there is at least one major tipping point in the earth's climate system that will be passed unless net carbon emissions reach zero and that the tipping point itself will be passed in much less than thirty years: "the intervention time left to prevent tipping could already have shrunk towards zero, whereas the reaction time to achieve net zero emissions is 30 years at best."[56] What I will continue to call "the date-of-last-opportunity" is, in the terminology of these scientists, the date at which the reaction time exceeds the intervention time, making the intervention impossible to carry out.

For some effects, the date-of-last-opportunity may be soon or even now, and sadly for some disasters, the date is already past. This is starkly and tragically embodied in the melting of the West Antarctic Ice Sheet (WAIS), mentioned in chapter 1. When I began to work on climate change in the early 1990s, the debates had already begun about what should be the target of mitigation efforts; and a rise in average global temperature of not more than 2°C, or 1.5°C, beyond the pre–Industrial Revolution average temperature had not yet been politically anointed as the official goal. In those days, many activists said roughly, "Whatever the target becomes exactly, we must at least be sure not to pass any catastrophic thresholds, like the threshold for the melting of the West Antarctic Ice Sheet." But, sure enough, in 2014 two different teams of scientists converged on the conclusion that this melting is now irreversible—this date-of-last-opportunity, the last chance to prevent this effect from becoming irreversible, has receded into history as a tragic lost opportunity.[57] The kilometers-deep WAIS—vast mountains of frozen water—is already melting and will continue to melt, and from this additional water alone ocean levels will ultimately rise around four or five meters above their current level. The process of melting will most likely take two or three centuries, but we have no good reason to doubt that it will definitely happen.

This will in itself be catastrophic for hundreds of millions of people who live in coastal cities and will be forced to abandon their homes. It will be at best deeply burdensome economically for their fellow citizens, who will bear the tax burden from the forced relocations and, in many cases, the influx into their communities of additional people who will need shelter and medical care and whose children will also need education, as well as the costs of building entire new cities as large as, say, New Orleans, Norfolk, Boston, Kolkata, Mumbai, and Shanghai, all of which will probably be substantially inundated within around 250 years, along with much of Bangladesh, Vietnam (the Mekong Delta will go under water early on), and numerous small island nations.

Other catastrophic thresholds, known and unknown, certain and uncertain, may loom in the mists of the not very distant future. As one epigraph indicates, leading authority Tom Lovejoy believes that the point of no return for the Amazon rainforest is near.[58] Direct empirical measurements suggest that we may be approaching the threshold for disruption of the Atlantic Meridional Overturning Circulation (AMOC).[59] Disruption of the AMOC would interrupt the European warming from the "Gulf Stream" and allow, among other things, countries in Europe to cool toward the temperature otherwise to be expected from their latitude. Various thresholds for the massive release of methane either from permafrost on land—releases are already occurring in Alaska and elsewhere in the Arctic[60]—or from ocean depths[61] are somewhere nearby in time, future or past. The direct effects of the methane would be brief compared to the effects of CO_2, but even a brief surge of methane could, for instance, drive average global temperature past the thresholds for the melting of major additional bodies of ice, such as, most notably, Greenland's ice sheet (larger than WAIS), which would produce further extremely long-term, but large, sea-level rise on top of that caused by the now-inevitable addition to the oceans of the water to be melted from the West Antarctic ice.[62]

The fact that whether profoundly important events—sometimes disastrous events—will occur centuries into the future will be determined by the choices made by the present generation is the result of a fundamental structural feature of time that undergirds the asymmetrical relationship between present generations and generations in the far future. This asymmetrical relationship explains further why the elements of our inherited phenomenology of agency that Scheffler describes as "the primacy of near effects over remote effects" and "the primacy of individual effects over group effects" are now profoundly misleading.[63]

Let us explore in a little more depth this last of the three reasons we saw in chapter 1 why action ought to be taken right now to limit destructive climate change: the historical contingency that we are in fact the last people with the opportunity to rein in the social practices of fossil-fuel combustion that are forcing the climate toward tipping points with possibly unbearable consequences. Our being the pivotal generation lies at the peak of a crescendo of reasons for our exceptional responsibility.

First, even when a threatening disaster is in no way one's own causal responsibility, if the disaster can be prevented at relatively modest cost, and the disaster is extreme, one ought to act to prevent it. This is the fundamental reasoning illustrated by the parable of the Good Samaritan.[64] The Samaritan is in no way causally responsible for the man's being in the ditch. But the man cannot get out of the ditch without help, and help is easy for any passerby to give. One need not be committed to any kind of maximizing requirement whatsoever, like a requirement that one should always perform the best available act in every situation, in order to believe simply that if on some particular occasions someone needs to be rescued from a bad situation, and one can do so with relatively little effort and cost, one ought to do it. Apart from any religious authority, this seems to be an element of common sense, although it is difficult to put one's finger on exactly why it seems so obvious.

I suspect that the common-sense rationale is something like a fundamental sense of proportion against a background assumption of human equality: it seems absurd that one should allow such a serious threat for someone else to persist when it would involve so little trouble for oneself to deal with it—refusal to act smacks of taking oneself ridiculously seriously by assigning so much significance to avoiding so little difficulty for oneself rather than relieving so much difficulty for another person. The challenges in making the transition from a carbon-burning energy regime to alternative sources of energy are on the whole relatively minor compared to the gargantuan consequences of persisting in the carbon energy-business-as-usual. For instance, those who would otherwise suffer unemployment––like coal miners, frackers, and their families—could be protected by arrangements for a just transition with fair burden-sharing.

Second, however, the climate change case is obviously in fact not a case like the Good Samaritan's in which a threatening disaster is in no way one's own causal responsibility, if one belongs to a nation with high cumulative carbon emissions that are continuing. This is one point at which the arguments about intergenerational justice cannot be entirely separated from the arguments about international justice. One's responsibility to deal with intergenerational climate threats is affected by one's general responsibility for the occurrence of climate change, and such causal or historical responsibility varies extensively across nations, very roughly according to the extent of their industrialization, as we saw in chapter 2. If one's nation has been contributing substantially to climate change and is at present still contributing, then one's nation's responsibility to contribute to the prevention of disasters like the melting of the Greenland ice sheets and the destabilization of the Amazon Forest by reducing emissions is a negative responsibility to cease undermining the climate system and to stop creating threats for members of future generations of all nations, not a positive responsibility to rescue future generations from a threat not of one's own making analogous to the Good Samaritan's.

In general, most people think that such negative responsibility deserves high priority, and that where one is actively contributing to damage and misery, one's duty is to stop it, even if stopping it involves considerable costs to oneself. One has no right to persist in creating threats to the lives and welfare of others, whoever they turn out to be. This consideration makes it reasonable that any one set of generations in a nation that has contributed substantially to climate change and is continuing to do so should bear a considerable burden, if necessary, to stop making matters worse. "Stopping" would involve immediately sharply reducing carbon emissions, which seems to have been precisely what would have been needed to prevent the now-inexorable melting of the WAIS.

Third, it seems inherently important that we are the ones with the last chance to avoid placing the seal of finality on the severity of the potential losses—we are those alive at the date-of-last-opportunity to prevent climate disasters.[65] We have just seen that in responding to a date-of-last-opportunity, the people in a nation with strong obligations to stop exacerbating climate change might be doing no more than, or only a little more than, they ought for the first two reasons to have been doing already in any case. Next, consider further whether beyond other such prior grounds for action and beyond the sheer magnitude of the potential catastrophe in, say, numbers of rights thwarted and numbers of lives blighted in the future from inaction now, it is additionally significant whether the potential agents are the ones alive at the date-of-last-opportunity. This situation would be as if the Samaritan not only had thrown the man into the ditch, but also somehow knew that he would be the last to pass that way before nightfall and that the stranger in the ditch could not make it through the night if not rescued now by him. It was now or never, and therefore it was he or no one to the rescue. Put differently, "The current generation is drinking in the last-chance saloon of climate change mitigation, and closing time is near."[66]

If in, say, 1992, leaders had realized that the generation then alive could still prevent the melting of the WAIS (if they could have), but that they were the very last ones, they not only should have expended a portion of effort and expense that might have seemed to be the "typical," or "fair," share of any "similar" generation, but they should have taken up a heavier burden precisely because they were the last to have the opportunity to prevent such a catastrophe for hundreds of millions of coastal dwellers.[67] In an important battle, one should do one's fair share every day, but if one could somehow know that the decisive day had arrived, and the outcome would be decided on this day, any ordinary person would think, I believe, that one should give even more on this day just because it was the last chance to win. Why? It is difficult to grasp any further reason, and perhaps any will be "one thought too many."[68] We may be at argumentative rock-bottom and have already specified the best reason: this is the last chance to avoid irretrievable loss.

Further Reasons? Past Sacrifice, Hope, and Continuity

But a first additional thought, among three, is that in a way the ultimate value of the sacrifices everyone else has already made on every other previous day depends on whether the sacrifices we make on this decisive day are adequate for success. If ours are adequate, all the sacrifices on all the other days will have led to victory; if not, the past sacrifices will have led nowhere. At this point in time, everything depends on the effort made on this climactic day by us. To some degree, at least, the present controls the past's value. Those of us fortunate enough to be living strongly influence not only where the future begins but also where the past ends. What we make of the present will deeply affect both. Perhaps out of solidarity one bears some responsibility to the good people of the past as well as to the people of the future to preserve what decent humans—past and present—have valued and have sacrificed to preserve.

Which past sacrifices?[69] I have been emphasizing the almost-total failure of most societies so far to rise to the challenge of climate change, and past sacrifices specifically for the sake of controlling climate change have been so meager that they do not seem to merit much consideration. But climate change threatens many of our most valuable institutions and practices—those constituting economically adequate, reasonably humane, culturally rich, and technologically advanced societies in the building of which some of our predecessors did indeed sacrifice and struggle. It is the sacrifices made in the creation of decent societies in which people can respect each other, cooperate willingly, and accept differences that it would be tragic to have come to an end because, for example, multiple mass migrations necessitated by climate-driven sea-level rise undermined minimal civility and sustained bitter conflict.

Suppose a boxer whose career was made possible by the sacrifices of others is engaged in a fifteen-round fight, and has won seven rounds and lost seven. Now everything depends on who wins the pivotal fifteenth round. It seems reasonable for the boxer to give it all he has got to give in this final round. And the generations alive at the date-of-last-opportunity to prevent a disaster seem to me to be analogously bound to make a maximum effort. One could think of them as unfortunate to be alive at this time and to have to do so much, or one can think of them as fortunate to live now and to be the ones who have the opportunity to accomplish so much for untold generations—to be the ones who make the difference.

Is there some fallacy or illusion here? Initially this reasoning seems disturbingly similar to the reasoning that makes it so difficult to withdraw from a losing war. If a society has been fighting a long but unsuccessful war, the society has what economists call substantial "sunk costs"—substantial investments have already been made and cannot be recovered. And one may think, if we withdraw from the conflict now, it will all have been for naught, but if by fighting a little longer, we can turn it around and win, it will all have been worthwhile. And because of such reasoning, pointless wars drag on and on, in the vain hope that persistence will turn the tide.

But the usual mistake is automatically to assume that simply fighting longer will in fact lead to winning. When one has been losing all along, will something automatically change? The problem is commonly this false premise, but not a fallacious inference. It is indeed sometimes the case that, *if* by fighting a little longer, one can turn it around and win, it *will* all have been worthwhile. The question is whether the if-condition will turn out to be true. The fighter who thinks, "I have taken a considerable beating so far, but I have won half the rounds, and if I win the last round, I will break the tie and win the fight," is not engaged in fallacious reasoning. He simply needs to make his premise true. He needs to be as sure as he can that he wins the last round. Maximum effort could indeed pay off. It seems similarly reasonable to expect the generations alive at the date-of-last-opportunity to prevent a disaster to do all they can, lest all previous efforts come to nothing, provided only that the further effort is not reliably known to be futile.

If one misses a genuine last opportunity to prevent a loss, the resultant regressive change becomes irreversible, and what is lost is irretrievable. I do not know any deeper reason why an irretrievable loss of a disastrous magnitude is vital to avoid, but one can add two further brief supportive conjectures, one about hope and one about continuity in valuing.

Before a change for the worse becomes irreversible, there is hope that it can be avoided, and the hope itself can be precious. Irreversibility brings the death of hope. The loss of hope represents, as Jonathan Schell wrote of extinction, a kind of second death: beyond the deaths of all the individuals, the death of the possibility of any more such individuals: "death cuts off life; extinction cuts off birth."[70] In the case of the WAIS, we have lost a major battle (before we had even begun to fight). We have lost any hope of avoiding a major rise in sea-level up every coast onto every continent and into scores of metropolises and countless villages, making hundreds of millions of lives miserable and perhaps desperate. This failure is deeply discouraging, and one thing that humanity cannot do without in the struggle to limit climate change is hope. We cannot succeed in preventing additional disasters unless we can lick our wounds and then return to the fight with refreshed hope and steeled determination to succeed. Rationally, of course, the frustration of one specific hope, even a major one, is no reason to abandon other separate hopes. And there are many more battles that we can win. But each lost battle is still a blow, and hope in itself is of value and to be preserved unless definitely ill-founded. Sometimes flagging hope can be revived by winning a new round with a final effort—another reason to try until the last.

Those of us who decades ago understood, or should have understood, the state of the planet have already failed to prevent substantial climate change. The climate is changing now, and will change more rapidly through many tomorrows. These changes will be especially difficult to adapt to for those to come who possess the fewest resources. But ahead lie many more important battles that can be won if enough people do not give up the fight. The outcomes of future struggles will determine crucially how much more severe and dangerous climate change becomes. One ought to keep hope alive for those generations of "strangers" who must make a life in the world we will have shaped, and one ought to prevent as much irreversible decline in the livability of the planet as possible so that they can enjoy lives of dignity, not drudgery.

Second, the preceding arguments appeal primarily to the humanity and potential dignity of those to come and imply that one ought to bequeath, at a very minimum, living conditions in which dignity is practically possible. Writing more recently about death, Samuel Scheffler has made the provocative suggestion that "the world of the future becomes, as it were, more like a party one had to leave early and less like a gathering of strangers" if one appreciates that "many of the things in our own lives that now matter to us would cease to do so or would come to matter less" if one were not confident that one will be succeeded by others who can and will value what we value.[71] This resonates with Hannah Arendt's reflection on the other of the two greatest threats to humanity besides climate change, nuclear weapons: "man can be courageous only as long as he knows he is survived by those who are like him, that he fulfills a role in something more permanent than himself, 'the enduring chronicle of mankind,' as Faulkner once put it."[72]

If either Scheffler's thesis about the dependence of our valuing upon subsequent valuing by others or Arendt's thesis about the dependence of courage on a confidence in survival is correct, it provides an at least partly, but by no means entirely, self-interested additional reason why we would be wise to leave a planet on which human life will not become nasty, brutish, and short, but will be lived in circumstances in which those who arrive at the party after we have departed will have the leisure and capacity to value and preserve much (at least) of what we value and try to preserve. This means at an absolute minimum that we must not allow conditions to deteriorate so far that the struggle for mere survival becomes all-consuming for even more people than it is today. This commitment in turn requires that we stringently limit severe negative changes that would be irreversible. We must not rest for as long as climate change remains unlimited and additional tipping points lie ahead. This is a further reason to act and to act even more energetically.[73]

Conclusion

What matters most, however, about a date-of-last-opportunity may be simply that inaction at such a time leads to irretrievable loss. Of course, in many cases no one can be sure that a particular date is a date-of-last-opportunity until long after the date has receded into the past. In order to have any effect, we must choose now—under uncertainty. This kind of uncertainty, however, tilts strongly toward action.[74] Either this is the date-of-last-opportunity for one or more disasters, or it is not.

Suppose it is not, but we choose to act in worthwhile ways like rapidly reducing carbon emissions. Then we will exert ourselves and incur expenses beyond our responsibilities toward people of the future. We will not help to save future generations from catastrophe but only to enrich the conditions of their lives to a degree that we have no duty to bring about. We will have morally "overachieved." But if the tasks we undertake are clearly not excessively burdensome for us, at worst we will have left a legacy for future generations that exceeds our responsibilities.[75] If moral overachievement is a "mistake," it seems like a good kind of mistake to make—and it seems a bit strange to think it is a mistake.

Or suppose it is. If in fact we are at a date-of-last-opportunity for one or more climate disasters, and we choose not to act, we will have allowed an avoidable disaster to engulf those who come after us. We will have done nothing while an irretrievable opportunity disappears. A disaster that we could have locked out will have become locked in to the climate system. If we miss the last opportunity, it is lost—forever.

An uncertainty between whether to risk putting in more worthwhile effort than we might have been required to—to overachieve—or to risk leaving the door open to a catastrophe that will reverberate through generations helpless to stop it—to fail to rescue untold millions from terrible fates—is not a reason for delay but a reason for action. Few "gambles" are so bearable on the downside and so promising on the upside, which is a spectacular opportunity—perhaps the last—to make an event that would be intolerably bad for whoever experiences it far less likely, if not impossible.

4

Are There Second Chances on Climate Change?

The six stages of climate denial are: It's not real. It's not us. It's not that bad. It's too expensive to fix. Aha, here's a great solution (that actually does nothing). And—oh no! Now it's too late. You really should have warned us earlier.[1]

Don't think of it as the warmest month of August in California in the last century, think of it as one of the coolest months of August in California in the next century.[2]

> And because I cannot apologize
> to the tree, to my own self I say, I am sorry.
> I am sorry I have been so reckless with your life.[3]

Fix It Later?

Perhaps today is, however, not in fact the date-of-last-opportunity. Even if what we fail to do today profoundly affects what happens in future, maybe what is done in future can nevertheless reverse the effects of our inaction today. First, to what extent can humans later undo the damage that we are doing now by our grossly inadequate response to climate change? And, second, how should any possibilities of later recovery or repairs influence our judgments in the present about the stringency and urgency of action called for by the rapidly worsening climate? My fundamental answer to the second question is that the stringency and urgency of action now ought to remain unaffected by any hopes and dreams of a later "fix." The answer to the first question and the relation of this answer to the answer to the second are complex, if fascinating, challenges—more complex than I realized when I first began to write about them.

My first attempt to answer these two questions directly was the controversial article "Climate Dreaming."[4] I believe that my basic caution in that 2017 piece against relaxing in the present in hope that people in the future can repair the damage we do now was well justified, even better justified than I knew then. The issues, however, are also more complicated than I understood then and require a more sophisticated treatment that I will attempt to provide here. In the interim, I have benefited especially from early and empirically well-informed philosophical critiques of my arguments by Dominic Lenzi[5] and from subsequent comprehensive overviews of the empirical literature.[6] Before tackling specific issues, we might remind ourselves of our own historical setting.

In 1992, many eminent national politicians from around the world, among them an impressive collection of heads of state including US President George H. W. Bush, gathered dramatically in Rio to adopt the Framework Convention on Climate Change and to promise effective steps to limit climate change. Those promises have been broken. Few steps of any significance have followed, and a crucial quarter of a century and more in which climate change could have been limited to a much lower level than it will now reach has been wasted while our supposed leaders have failed to lead on—or even pay attention to—the most important threat facing civilization.[7] It is little wonder that the generation of Greta Thunberg is angry that opportunities that could easily have been taken decades before their births were repeatedly frittered away while the fossil-fuel regime entrenched and expanded business-as-usual, and the rate of carbon emissions continued to accelerate.

This quarter of a century of political failure and corporate deceit and greed matters most fundamentally, as we have already seen, because of the cumulative carbon budget. A virtually linear relationship holds between the average global temperature of the earth and the total atmospheric accumulation of CO_2 since the date at which the combustion of carbon-based sources of energy—coal, oil, and natural gas—began to change the composition of the earth's atmosphere. David Wallace-Wells summarizes with characteristic vividness: "We have done as much damage to the fate of the planet and its ability to sustain human life and civilization since Al Gore published his first book on climate [in 1992] than in all the centuries—all the millennia—that came before."[8] While this is not quite precisely correct—cumulative global emissions of CO_2 were 847.78 billion tonnes in 1992 and 1.58 trillion tonnes in 2017[9]—it certainly portrays the general situation accurately. Between 1992 and 2017, the total accumulation of CO_2 in the earth's atmosphere had almost doubled as a result of the betrayal of the political pledges made in Rio. Elizabeth Kohlbert makes a similarly trenchant point about the fact that the global rate of emissions has been sharply accelerating every decade (and almost every year): "In the past ten years, more CO_2 was emitted than in all of human history up to the election of J. F. K. [in 1960]."[10] Or, another way to think of it, "burning the first trillion [barrels of oil] took about 130 years, but we went through the second trillion in only twenty-two years."[11] Or, "the last 30 years have been the golden age of fossil fuels. Far more fossil fuels have been burnt in the past 30 years than in the entire nineteenth century."[12] The Greenland ice sheet alone has lost 4 trillion tons of ice since 1992.[13] You get the trend.

The 2019 edition of *The Production Gap* provided this summary: "In 2019, as climate impacts intensified, global fossil-fuel combustion was at an all-time high. Coal, oil, and natural gas remain the world's dominant sources of energy, accounting for 81% of total primary energy supply. These fuels are, by far, the largest contributor to global climate change, accounting for over 75% of global GHG emissions . . . and close to 90% of all carbon dioxide (CO_2) emissions."[14] In spite of the fact that fossil-fuel emissions must reach net zero in only a few more years if average global temperature is to stabilize below a rise of more than 1.5°C, and in only a few years later to stabilize below a rise of more than 2°C, global sales and combustion of coal, oil, and natural gas all continued to rise through 2019.[15]

One crucial factor is that while the use of clean fuels like solar and wind has increased surprisingly rapidly, this energy is mostly consumed on top of fossil-fuel energy rather than replacing it.[16] Total energy consumption steadily rises, and the alternative energy is largely an addition to, not a substitute for, fossil-fuel energy. Successful policies will need, as we will see in chapter 5, to confront the fossil-fuel regime head on and aggressively drive down the extraction and sale of coal, oil, and natural gas. The current level of mitigation of fossil-fuel emissions is not remotely ambitious enough.

In fact, decades of past failure to confront climate change have now allowed such a large amount of additional emissions of fossil fuels that it is unavoidable that the total atmospheric accumulation of CO_2 will exceed the carbon budget for any tolerable rise in average global temperature, at least temporarily. We must face this terrible result of our political failure unblinkingly. In addition to everything else that must be done, significant amounts of the CO_2 injected into the atmosphere in recent decades by human energy consumption must be removed. We confront the necessity, then, for what the climate modelers initially called "negative emissions." To begin, we must get past some clumsy labels and possible terminological confusion, concerning three points.

First, "negative emissions" refers to the reversal of emissions, that is, the extraction from the atmosphere of what has previously been injected into it by so-called negative emission technologies (NETs). This phrase is clumsily nonintuitive terminology that is, I suspect, the product of hasty jargon-coining by a number-cruncher fixated on the fact that in this case the emissions would have a negative sign—this would be an amount by which emissions were coming down rather than going up. This jargon is rather like calling the opposite of inhalation not "exhalation," but "negative inhalation," or referring to fasting as "negative eating." To some extent, the label of "negative emissions" is more recently being replaced by "carbon dioxide removal" (CDR), which is certainly more intuitive and which is the phrase I will adopt here. CDR has the added advantage of sounding like operations performed on nature by human technologies, which is indeed what it would be.

Second, the other bit of complementary new jargon is "overshoot": the amount of fossil-fuel emissions that exceed any given cumulative carbon budget—the budget for any given amount of rise in average global temperature—is now referred to as the overshoot relative to that budget. The purpose of CDR, then, is, in the jargon, to eliminate emissions overshoots and bring the actual atmospheric concentration back inside a particular cumulative carbon budget. When the climate modelers say that the temperature rise can be "kept" to a certain level, such as 1.5°C, what they usually mean is considerably less straightforward: that the temperature can be expected to stabilize at that level by around 2100 if any midcentury overshoot in emissions has meanwhile been reversed decisively enough. The idea is that the temperature as well as the level of cumulative emissions will unavoidably, as a consequence of political failure until now, exceed our targets in the middle decades of the century, but if it is feasible to eliminate the overshoot in emissions soon enough, the temperature will drop back down to target level by around 2100.

Third, mitigation is more "ambitious" insofar as it contributes to reaching net zero emissions of CO_2 globally at an earlier date— and therefore at a lower level of cumulative atmospheric accumulation of CO_2, or within a smaller carbon budget.[17] The "ambition" of a national commitment to mitigation, therefore, refers not only to the extent of the decarbonization to be conducted within a nation's own borders, but also to the extent of the same nation's financial support for decarbonization to be carried out elsewhere, which together have been aptly called its "dual obligations" for mitigation.[18]

A nation's responsibility for mitigation may exceed its economically sensible capacity to reduce its own domestic emissions, with what is economically sensible calculated at the global level; in that case, it ought to provide financial or technological transfers that enable the reduction of emissions in other nations. "Ambition" in the sense discussed here, then, combines a nation's Nationally Determined Contributions (NDCs) regarding reduction of its internal emissions with its financial and technological commitments in support of reduction of emissions elsewhere. How the ambition of individual nations should be divided between internal emissions reductions and external emissions reductions may be partly dependent on global-level efficiency considerations, but its minimum total effort should reflect both its national historical responsibility, as explained in chapter 2, and its responsibilities of international and intergenerational justice, as we just saw in chapter 3.

Contrasting Purposes of CO_2 Removal

An influential 2016 study noted that the database of the Fifth Assessment Report by the Intergovernmental Panel on Climate Change (IPCC), released in 2014, "includes 116 scenarios that are consistent with a >66% probability of limiting warming below 2°C (that is, with atmospheric concentration levels of 430–480 ppm CO_2eq in 2100). Of these, 101 (87%) apply global NETs in the second half of this century."[19] So 87% of the IPCC's scenarios for the 2014 report that allow the target of 2°C to be achieved rely on NETs or, as I think it is now better to call them, CDR. The Special Report on 1.5 says, "All pathways that limit global warming to 1.5°C with limited or no overshoot project the use of carbon dioxide removal (CDR) on the order of 100–1000 $GtCO_2$ over the

21st century."[20] This heavy dependence on CDR for stabilizing at nondisastrous temperatures by 2100 is not widely appreciated by the general public.

CDR can be intended to accomplish any one of three radically different purposes: (1) enhancing current mitigation, (2) remedying insufficient past mitigation, and (3) rescuing fossil-fuel companies' stranded assets. I will call these three types respectively "enhancement CDR," "remedial CDR," and "asset-rescue CDR."[21] Asset-rescue CDR is in the interest of no one except those whose wealth is tied up in reserves of, and infrastructure for, coal, oil, and natural gas, which at present still includes, besides the companies themselves, socially important players like large pension funds that persist in clinging to stock in the companies. Pension funds ought to divest these holdings rapidly for reasons of both self-interest, since fossil-fuel corporations are headed into decline,[22] and morality, rather than expecting anyone other than the executives and owners of the companies to support asset-rescue CDR, which I will discuss no further.[23]

The first two purposes, enhancement CDR and remedial CDR, are distinguished as follows in *Global Warming of 1.5 °C*: "CDR can be used in two ways in mitigation pathways: (i) to move more rapidly towards the point of carbon neutrality and maintain it afterwards in order to stabilize global mean temperature rise, and (ii) to produce net negative CO_2 emissions, drawing down anthropogenic CO_2 in the atmosphere in order to decline global mean temperature after an overshoot peak."[24] Enhancement CDR is essentially preventative: ambitious reductions in current emissions can be supplemented and augmented by the concurrent removal of previous emissions. At any given time, it may make no difference to the extent of climate change whether one less ton of CO_2 is emitted or one more ton of previously emitted CO_2 is extracted from the atmosphere. What matters, as we have seen, is the resulting cumulative total of CO_2 in the atmosphere. If, for any given expenditure, one could reduce the subsequent atmospheric total more by CDR than one could by emissions reductions, then it would be more efficient to carry out the CDR rather than the mitigation, other things being equal, which, as we will see further just below, they often will not be. At present, however, mitigation is far cheaper than CDR. For one thing, any carbon removed would have to be sequestered securely for as long as ten thousand years, and the necessary infrastructure to do this at any significant scale is not in place or even concretely planned.

Remedial CDR is essentially corrective: if mitigation (plus any enhancement CDR) had been ambitious enough and the atmospheric concentration of CO_2 had not exceeded the carbon budget for a tolerable average global temperature—in other words, if net zero CO_2 emissions had been achieved at an early enough date to keep the temperature rise sufficiently small—no CDR for the second purpose of remedying overshoot would have been needed. But when the now-inevitable overshoot in CO_2 emissions has occurred, one or more forms of CDR will be needed to reduce the cumulative atmospheric concentration of CO_2 back down to the level at which net zero CO_2 emissions ought to have been attained. As I have already noted, action to control climate change has been so pathetic over recent decades that it is now virtually impossible to prevent an overshoot relative to the carbon budget for any "tolerable" rise in temperature. The only open question is how serious the overshoot will be and how much extra avoidable damage to society and nature an overshoot of that extent will do. Preventing yet further damage will of course then be crucial, and substantial CDR should if possible be accomplished as long as it is doing more good than harm.[25]

What I was objecting to in "Climate Dreaming"—and important articles by others have also objected to[26]—was not CDR carried out for either of the above two purposes, but less ambitious reductions in emissions in the present on the excuse and groundless assumption that the emissions not forgone now can unproblematically be removed later. I was criticizing the folly of imagining that the desperate need for immediate ambitious reductions in emissions (augmented if possible and efficient by enhancement

CDR now) had been obviated by the apparent possibilities for removing CO_2 later through remedial CDR. Any such glib confidence in "fixing it later" with remedial CDR is deeply misguided for three major reasons that I will begin to spell out shortly. The objection is not to remedial CDR as such, which will be a good thing at some scale to alleviate unprevented overshoot. The objection is to allowing the awareness of hypothetical remedial CDR later to provide a pretext for actual failures to mitigate as much and as rapidly as possible now, causing the later overshoot to be larger than it would have been with more ambitious reductions now.

The problem is in part—but certainly not entirely, as we will see in the remainder of this chapter—what lawyers tend to call "moral hazard": the supposed fact (accurate or not) that later recovery is possible from damage done by less care now tempts one to show less care now because one believes one has grounds for confidence about avoiding paying the price for the present carelessness. The less aggressive mitigation now, not the dreamed-of remedial CDR later, is the fundamental mistake, although much depends on when, where, how much of, and what kind of remedial CDR is hoped for—and actually accomplished.

Some remedial CDR will be necessary, and additional enhancement CDR soon could lessen the amount of remedial CDR that becomes necessary: "Carbon removal does not avoid the need for rapid emissions cuts, which remain essential. But if removal can work at the required scale with acceptable impacts, using it in parallel with deep emissions cuts could substantially reduce climate risks. CDR at large scale could also make climate targets achievable that are out of reach with emissions cuts alone."[27]

Small-scale CDR using any of several techniques can on balance be helpful, and the worst problems emerge only when one technology for CDR becomes large scale—if, for example, it requires extensive land, water, or energy more urgently needed to grow food instead. The crucial issues for now are how much and what kinds of remedial CDR will have to be performed for the purpose of correcting past lack of action; and we must avoid leaving future people to try to carry out larger-scale remedial CDR than would have been needed but for our lack of effort at and investment in mitigation now, including enhancement CDR (up to some scale).[28] Above all, the possibility of CDR later must not be allowed to motivate persistence in half-hearted mitigation now in the utterly vain and totally groundless hope that CO_2 removed later would be equivalent to CO_2 emissions avoided now. It would not. The main purpose of this chapter is to show why not, that is, to explain why removal of CO_2 later—especially attempting removal at large scale—is not as good as either preventing CO_2 from being injected into the atmosphere now or removing it by enhancement CDR now. Three independent reasons converge to make a powerful case against complacency now, grounded in hoped-for remedial CDR later.

Scaling Up in Time?

The first reason why removal later may not be as good as preventing the problem from growing so large is that CDR at the scale that will be necessary if ambitious mitigation is not urgently implemented now may simply not be possible in time, given how little has been invested in developing CDR until now. While this contingency may well be fatal to hopes that CDR can accomplish the second, remedial purpose, it is obviously farthest outside my competence as a political philosopher. Nevertheless, since it is a potentially decisive challenge, it must at least be mentioned superficially.

One of the conclusions of a major 2018 survey of techniques for CDR was that "a substantial gap exists between the upscaling and rapid diffusion of NETs implied in scenarios and progress in actual innovation and deployment. If NETs are required at the scales currently discussed, the resulting urgency of implementation is currently neither reflected in science nor policy."[29] The IPCC's Special Report in 2015 found that "most CDR technologies remain largely unproven to date and raise substantial concerns about adverse side-effects on environmental and social sustainability."[30]

In general, there is no mystery about the extraction of CO_2 from the atmosphere. This is how trees and other plants live: they extract CO_2 and combine it with other elements through photosynthesis. Because there is substantially more vegetation (because more land) in the Northern Hemisphere than in the Southern, total global atmospheric CO_2 rises over the Northern winter when most Northern vegetation tends to be dormant, begins to drop in Northern spring as vegetation comes back into action, and reaches its annual low point at the end of Northern summer when photosynthesis has completed its maximum work for the year. Some people describe these remarkable seasonal falls and rises in CO_2 as life on the planet as a whole inhaling and exhaling—breathing in and out CO_2. Of course, the vegetation in the oceans, the water in the oceans, the soils on land, and other elements of our planet do their own complex "breathing" of CO_2 in ways that I do not pretend to understand very well but that need not concern us here.[31]

The fact that trees (and other plants including aquatic ones) carry out a net removal of CO_2 from the atmosphere is obviously why tree planting can help to control climate change—some of the carbon taken in by a tree is kept and used and thus stored in the wood, safely out of the atmosphere at least temporarily until the wood is burned or decomposed. This is also why the ferocious megafires in California and Australia that were made more likely by climate change are themselves positive feedbacks that make climate change worse still by driving up the atmospheric carbon level. And it is why the massive fire-assault on the Amazon Forest by beef-raising and soy-growing Brazilian agribusinesses that grab more land without having to pay for it (and murder indigenous people who try to defend their forest homes), and that are cheered on by President Bolsonaro in the name of wildly implausible conceptions of national sovereignty and of economic development, are so damaging to all other living beings on the planet.[32]

Our focus in this chapter is on the question of the proper role in policy on climate change for technologies of CDR, given that decades of failure to control emissions has made overshoot of most reasonable carbon budgets impossible to prevent at this late date. Technologies for CDR include afforestation, reforestation, soil carbon sequestration, bioenergy with carbon capture and storage (BECCS), direct air carbon capture and storage, and others.[33]

Soil carbon sequestration, for example, makes soil more productive, thus partly paying for itself; is easy to implement; and is a relatively uncontroversially good way to capture some carbon. Afforestation and reforestation (and of course the even more vital prevention of deforestation) play positive roles up to the point at which the trees compete for land and water with efficient and essential forms of agriculture, especially growing vegetables for direct human consumption. Grazing beef on pasture converted from tropical forest and growing soybeans to feed Chinese pigs on land converted from tropical forest,[34] as is increasingly happening in the Brazilian Amazon and in the Cerrado, are flagrantly extreme cases of grossly inefficient, totally unnecessary, and wildly destructive agriculture. Many other forms of meat production are also highly wasteful of and polluting of basic resources like land and water. Until agriculture started being conducted according to industrial principles by agribusinesses, farms and forests coexisted happily in many different combinations in different parts of the planet.

Plainly, the size of human populations is another major factor in how much land and water must be used even for the most land-efficient forms of food growing. Basically, trees can help the climate until they reach the scale at which they interfere with necessary measures to feed people, the extent of which depends on how many people there are and what diet they eat. The further complications about farms and forests raise many vital issues, but we must settle for this superficial and broad glance here so that we can focus on the type of CDR that has been most strongly endorsed in climate policy: BECCS.

In theory, BECCS would have the great merit that it actually produces net energy—this is the bioenergy part—while removing carbon from the atmosphere in two steps, initially through the photosynthesis involved in the growth of the biological material that is the feedstock for the combustion, and then through the CCS process that is supposed to capture most of the emissions from the combustion of the biomass that produces the energy. But BECCS has high land- and water-use intensity. The most widely cited study of BECCS found that, in order to remove 3.3 $GtCO_2$ per year (3.3 Gt is 3.3 billion metric tons) from the atmosphere, the growth of conventional feedstocks for the bioenergy required would use approximately "3% of the freshwater currently appropriated for human use" and "a land area of approximately 380–700 Mha in 2100," which is one to two times the size of India![35] Using this much water and land to grow feedstocks for bioenergy would have serious repercussions for supplies and prices of food, and 3.3 $GtCO_2$ per year is a small fraction of the remedial CDR that is likely to be needed. As Bellamy and Geden observe, "The scenario pathways in the [IPCC 2015 special] report that limit the warming to 1.5°C rely heavily on CDR, with a median deployment of 730 $GtCO_2$ removed over the course of the twenty-first century. The latest estimates put annual anthropogenic CO_2 emissions at 42 $GtCO_2$, so the envisaged level of CDR is on the order of almost 20 years of today's emissions."[36] Removing 3.3 $GtCO_2$ per year would constitute canceling only about 8% of the current annual emissions of 42 $GtCO_2$ per year.

Now, all sorts of empirical complications are present, but we cannot possibly explore them here. For example, although one obvious reason to worry about large-scale BECCS is the massive amount of land and water it could use, bioenergy is not used only for removal of previous emissions, that is, not only as a kind of CDR in the form of BECCS. Bioenergy is also used to reduce emissions, that is, as a kind of mitigation in the form of transport fuel, like the highly subsidized ethanol beloved by politicians from the Middle West, substituting for fossil fuel. Less bioenergy for

CDR would not necessarily mean less bioenergy overall—indeed, it could mean more bioenergy for emissions reduction in order to reduce the need for emissions removal.[37] Further, a great deal of research is being done on nonconventional BECCS feedstocks that might not require as much land and water as needed by conventional feedstocks like Iowa corn and Brazilian and Indian sugar.

On the other hand, however the feedstocks are grown, BECCS requires two monumental kinds of transport networks: the gigantic amounts of feedstock must constantly be transported to the facility in which the combustion will occur to keep it supplied with fuel, and the CO_2 captured by the CCS must be transported to a geological formation in which it can be securely stored for at least ten thousand years. Moving the feedstock would require vast numbers of trucks, trains, or other vehicles, and moving the captured CO_2 would require vast systems of pipelines or tanker cars for liquified CO_2 (liquified to make it less volatile for transport and storage). Little of this infrastructure exists, and few are seriously investing in it, least of all the fossil-fuel companies whose product creates the problem. Some BECCS facilities could be located near a geological reservoir (like empty oil fields under the North Sea), and some could be located near the source of the feedstock, but there is no reason to think that all three could often be in the same place. If, for instance, the CO_2 is going to be pumped under the North Sea into formations now emptied of fossil fuel, are the feedstocks going to be grown in the Scottish Highlands so that the BECCS facility can be on the Scottish coast? Not likely.

In any case, all we need to understand for our purposes is that the feasibility and affordability of BECCS in particular and CDR in general are open questions, especially at the large scale that is extremely likely to be needed. The difficulties already raised in the literature have been producing consideration of a larger range of options, including "portfolios of multiple NETs, each deployed at modest scales."[38] Soil sequestration and reforestation at considerable scale, and BECCS at modest scale at least, for example, can surely help,[39] but they cannot possibly produce anything like the median deployment of 730 GtCO$_2$ removed that the scenarios modeled require. It is misguided simply to dream that however lackadaisical and underfunded our efforts at controlling climate change continue to be, future generations can simply repair everything later with CDR. And there is *much* more wrong with leaving climate change for future people to bring under control than the simple but important fact that it may in reality be possible to fix only so much later, as we will see in the next two parts of this chapter.

Meanwhile, we can note in passing that some proponents of solar radiation management (SRM) will see the difficulties in implementing massive CDR as reasons why we ought to pursue SRM instead of, or in addition to, CDR. I consider SRM a dangerous, unpredictable option that should be implemented, if ever, only as a last resort in desperate circumstances resulting from decisive failures in mitigation and CDR. The issues SRM raises are deep and complex, and a proper discussion of them would require an additional chapter for which I do not have space here.[40]

Bequeathing Risks

We now turn to a normative analysis of the relation between the level of the present-day ambition to stop contributing to climate change and the risk of danger for future generations—between present inaction and future danger. We explore the structure of the risks created by every decision about how ambitiously to mitigate the primary factor forcing climate change, CO$_2$, and thereby reveal the deepest objection to the current half-hearted and half-baked mitigation. The two main theses in this section of the chapter are (1) that, in general, all decisions in the present about the degree of ambition for emissions mitigation are unavoidably also decisions about how to distribute risk across generations; and (2) that, more specifically, the less ambitious the mitigation is, the more inherently objectionable the resulting intergenerational risk distribution is.[41] One of the most tempting ways by which a country's policy-makers can persuade themselves that less than a reasonable amount of mitigation now is permissible is by assuming that remedial CDR later would be just as good as more mitigation now—by gambling that remedial CDR will be fully equivalent to ambitious mitigation (including any enhancement CDR) now. This is not a reasonable gamble, as I will now try to show. The following normative structural analysis of risk thus has strong implications for the standards of assessment by which to judge national implementation of the 2015 Paris Agreement, especially the extent of the ratcheting up of each nation's NDCs as mandated by the Agreement.

Every risk, as Hermansson and Hansson note, involves "three roles, namely the risk exposed, the decision-maker and the beneficiary."[42] The three roles may be occupied respectively by one, two, or three parties. How the three roles are allocated among the parties involved gives the risk what I am calling its "structure." In the simplest case, all three roles are occupied by the same person who chooses to gamble and receives the benefits or suffers the losses of the gamble. If, instead, for example, the decision-maker is one party and a single other person is both the intended beneficiary and the risk-exposed, the structure of the choice may be paternalistic.[43] For instance, the legislature requires motorcycle helmets, and cyclists benefit from greater safety and suffer the loss of the thrill of greater danger.

Or, to see a different structure, take the typical choice by a firm to externalize the environmental costs of its operations. The firm is the decision-maker about how much to pollute, and it is itself the beneficiary of the reduced costs relative to controlling the pollution, while third parties bear the costs and suffer the harms created by the pollution. Other people are stuck with the costs of dealing with the pollution in order to reduce production costs for the firm. This structure constitutes corporate exploitation of the public. We saw in chapter 2 that externalization at the national level—sovereign externalization—uses the remainder of the planet for waste disposal.

John Rawls focused attention on a simple situation for choice between two alternatives, which I have modified in the following in important respects, partly following Stephen Gardiner.[44] Alternative A has a high probability of producing a satisfactory outcome. Alternative B has three specific features: (1) it might produce trivial relative gains compared to A, (2) it might produce significant relative losses compared to A, and (3) attaining knowledge of the probability that it might produce the trivial gains rather than the significant losses "is impossible, or at best extremely insecure."[45] Manifestly, choosing alternative B would be paradigmatically imprudent—alternative B is a very bad gamble, risking a substantial loss for the sake of a trivial gain in circumstances of uncertainty.

Our fundamental choice with regard to climate change is about how ambitious to make mitigation now—about when to aim to bring carbon emissions to net zero globally and leave behind the era of fossil fuels. While a few of the initial NDCs made at Paris 2015 may have been adequate tentative first steps—"first steps" coming three decades after they were promised!—most of the NDCs ranged in ambition from the merely perfunctory (e.g. the American) to the paltry (e.g. the Australian). To settle thus for unambitious mitigation now is to gamble that remedial CDR later by subsequent generations can rescue future people from what would otherwise result from our present complacency.

More ambitious mitigation will clearly impose some costs on some segments of the current generations. Nevertheless, if any sacrifices necessary for the sake of more ambitious mitigation are shared fairly—if contemporaries put into place fair distributive mechanisms—no life in current generations that is satisfactory now need become unsatisfactory because of the costs of ambitious mitigation. Mitigation through a transition to alternative sources of energy would itself make many people much better off and some people worse off. But those whose well-being would decline can be protected against falling below a satisfactory level by fair mechanisms of transfer from those whose well-being improves. For example, coal miners who become unemployed can be retrained for much safer and better jobs, or, if they are too old for retraining, be provided with unemployment benefits sufficient to support their families. Actually arranging for fairness in the distribution of the mitigation costs is a crucial assumption, of course. It is not enough that the transfers are possible; they must be institutionalized.

On the assumption just sketched, we can compare the situation of the current generation with the situation of the individual discussed by Rawls. Like that individual, the current generation must also choose between two options, in this case two tendencies in mitigation: more ambitious (prompt, large emissions reductions now, perhaps augmented by enhancement CDR) and less unambitious (reliance on the prospect that future people will be able to recover from the path we leave them on by using remedial CDR). More ambitious mitigation by the current generation with fair institutions could allow satisfactory outcomes for everyone now whose life would be satisfactory with less ambitious mitigation. On the other hand, to the extent that mitigation was less ambitious, it would tend to have three features: (1) it might produce minor relative gains for some people compared to more ambitious mitigation by avoiding some expense, effort, and disruption; (2) it might produce significant relative losses for some people compared to more ambitious mitigation by postponing the date of net zero carbon and thereby allowing climate change to become worse than it would be if mitigation were more ambitious early on, and (3) knowledge of the probability that it might produce the trivial gains but not the significant losses "is impossible, or at best extremely insecure."

David Weisbach has described what I take to be the same gamble between more and less ambitious mitigation in the present as follows: "uncertainty about the effects of climate change strengthens these conclusions because the uncertainty is not symmetric: if we do nothing or act too slowly, the bad cases if things turn out worse than expected are far worse than the good cases are good if things turn out better than expected."[46] The gamble on less ambitious mitigation, then, is clearly at least as bad a gamble as the paradigm imprudent gamble that could have been taken by the individual described at the start following Rawls. In the mitigation gamble, we impose on future generations a gamble with a structure we could never reasonably choose for ourselves. And the less ambitious the mitigation, the worse the gamble, other things being equal, because the longer the period during which climate change can worsen, or even become catastrophic.

The alternatives in the Rawlsian case and the alternatives in the climate case are not perfectly parallel. Rawls simply compares two fixed alternatives. Mitigation involves degrees: relatively more ambitious and relatively less. This difference is not problematic, however, because each case still involves a choice between two options. Two other monumentally morally significant differences between the structures of the two gambles are evident, however. In the Rawlsian case, first, the possible gains or possible losses would go to the same individual; and, second, the recipient of the possible gains and losses is the person who decides whether to take the bad gamble. In contrast, in the mitigation case, the possible gains and possible losses would go respectively to different generations: the *gains* would go to the *current* generation, who would avoid whatever expense, effort, and disruption would be involved in speeding up the energy transition through more ambitious mitigation and enhancement CDR, but the *losses* would go to *future* generations, who would suffer whatever consequences resulted from worsened climate change occurring during the additional time prior to zero carbon emissions necessitated by less ambitious mitigation.

So, first, the mitigation gamble is this: possible gains for us, possible losses for them. And, second, the decision whether to make the mitigation gamble is entirely in the hands of the generation who can only gain, the *current* generation—the potential winners have all the *power*. The *future* generations who can only lose have no say whatsoever—the potential losers have *no power*. These two differences combined turn what is inherently a bad gamble into a reckless gamble in which for the sake of possible small gains for us the current generation imposes upon future generations whatever losses might come from climate change worsened by casual mitigation now. And this objectionable structure would be present even if we had no reason to think that those losses for future generations were especially likely or particularly serious. They are both.

In the instance of the choice between less ambitious and more ambitious mitigation, the three roles that every risk involves are occupied by only two parties. One party, the current generation, is both decision-maker and potential beneficiary. Future generations are exclusively the risk-exposed. This is why *decisions on mitigation in the present are decisions on risk distribution across generations.* This structure, in which one party decides and potentially benefits itself, while another party is risk-exposed, is exploitative. The decision-maker sticks the risk-exposed with the potential costs of potential benefit to the decision-maker itself. This exploitative structure is exactly the same as the structure of the choice by a firm to externalize the costs of its pollution: less ambitious mitigation "externalizes" risks from current generations to future generations in order to save us modest cost and inconvenience. This is the temporal analogue of sovereign externalization, dumping risks across time rather than across space. This can aptly be designated "temporal externalization." *Temporal externalization* unjustly treats future people in a manner parallel to the way that sovereign externalization unjustly treats (noncompatriot) contemporaries.

We can, therefore, make three normative judgements about any intergenerational gamble that chooses less ambitious mitigation in the present, even before we know anything about how likely or how serious the potential losses for future generations are. First, what would be a bad gamble for an individual to take upon herself could be, at the very best, an equally bad gamble for one generation to impose on other generations.[47] And, the less ambitious the mitigation, the worse the gamble. The intergenerational case, however, is far worse than merely being a reckless gamble.

For, second, a widely shared normative principle is that while one is at liberty, with some qualifications, to choose for oneself to run whatever risks one likes, one is not at liberty to impose the same risks on others without their consent. I am free to take a bad gamble if I wish, but I am not free to impose a bad gamble on others who have no voice in the situation. I am instead bound to exercise due care on behalf of others. One need not be risk averse on one's own behalf, but one has a duty of care to be strongly risk averse toward others. Generations that for the chance of small benefits for themselves choose less ambitious mitigation that risks major harms and costs for others fail in this duty of care. The less ambitious the mitigation, the greater the failure.

Third, I am all the more not free to impose on others a bad gamble on which only I can gain and they can only lose—this would constitute my using them strictly as a means to my own ends with no regard whatsoever for their purposes. Well beyond a simple failure of due care, this constitutes pure exploitation and a failure to exhibit minimal respect toward other people and their purposes. The possible effects of the current foot-dragging on mitigation are certainly nothing to impose on your descendants. This much is clear simply from the structure of the choice about risks.

As David Weisbach has noted to me, this gamble comes exceedingly close to heads I win, tails you lose.[48] Most harms from the current failure of mitigation will descend upon people in the future. Some benefits may accrue to people in the present, although it is seriously debatable how great the benefits are for anyone other than those who are invested in fossil-fuel assets and its dependent infrastructure, like pipelines, oil tankers, and freight railroads. This is not the place to pursue the economic considerations, but it is worth noting in passing one of the economic arguments based in an only moderately long view of self-interest in favor of ambitious current mitigation:

> Tackling the climate crisis will require the world's largest ever peacetime investment. Historically low interest rates mean there has never been a better time to make it. Interest rates affect the entire economy, but are particularly important for renewables because the cost of borrowing has an outsized influence on their competitiveness. Fuel costs are essentially zero for green power—sun, wind and water are free. That means capital expenditure is the biggest component of the average cost of producing renewable electricity. In other words, renewables' costs tend to be front-loaded, requiring upfront borrowing, and they deliver savings over time. The same is true of many technologies we need to reduce global emissions. It costs more to build well-insulated houses, but less to run them. . . . The cost advantage of renewables is startling.[49]

In addition, the need for extraordinary expenditures in order for economies to recover from lock-downs because of the coronavirus create a uniquely propitious time to "build back better."

It may appear that the main argument above slides improperly from a case of prudence to a case of morality. The Rawlsian individual is deciding between two options on the basis of her rational self-interest; it is not prudent for her to choose alternative B. The current generation is deciding between its own (comparatively minor, if not merely apparent) interest and the (potentially substantial) interest of future generations; it is not morally permissible, I am claiming, for the current generation to choose its own minor gain at risk of imposing major damage on future generations. The first Rawlsian choice is indeed a matter of prudence, and the second, a matter of morality. But the point is precisely that the structure that makes one of the two options in the individual choice imprudent, where the interest of only one person is at stake, makes one of the two options in the generational choice immoral, where the interests of additional generations are at stake and ought to be treated with care. Future humans, whoever the individuals turn out to be, deserve respect.

The Reversible and the Irreversible

The incurable failing of all remedial CDR is the ratchet effect of passing tipping points for the climate during the period of "overshoot." If some types of remedial CDR can be made to work at a global scale, then the atmospheric accumulation of CO_2 can go down as well as up. But if that accumulation, even if it ultimately turns out to have been transitory itself, meanwhile drives some crucial factor affecting the climate past a point of no return, that change in climate will be permanent in spite of the fact that its cause was temporary. *Temporary changes can produce permanent effects.* This is the greatest danger from the current limp approach to mitigation by many incumbent politicians.

For example, the cryogenic scientists who have concluded that the West Antarctic Ice Sheet probably is now irreversibly melting think that the reason why the melting is irreversible is that the crucial Thwaites Glacier is a marine-based ice sheet: it rests on land, but that land is underwater.[50] And as the land on which the ice rests extends back inland away from the ocean, it slopes downward under the ice sheet. The forward edge of the glacier at its grounding line—the last and highest point at which the ice sheet rests on land—is in contact with ocean water. This opens the glacier to "basal melt by ocean heat flux"—warm ocean water can melt the front of the bottom edge of the ice sheet.[51] If, as is the case with Thwaites Glacier, the land on which the ice rests slopes downward as it moves inland, the melting water flows downhill, reducing the friction with the land under the ice sheet and allowing the ice to slide more rapidly into the sea.[52] No one has been able to see anything remotely feasible that could stop this process of enhanced melting now underway and accelerating.

The function of Thwaites Glacier in the overall dynamics of WAIS is often analogized to the cork in a bottle: Thwaites is positioned now blocking a passage with elevated sides so that it holds back the remainder of WAIS from sliding more rapidly toward the sea. The same role of cork-in-the-bottle for the East Antarctic Ice Sheet (EAIS) is played by Totten Glacier. Totten too is a marine-based ice sheet that several lines of evidence now suggest is also susceptible to basal melt by ocean heat flux.[53] No one is claiming that the melting of EAIS too has become irreversible already—on the contrary, the point is that this may depend precisely on what we do in the present and immediate future. Consider, then, the following scenario; this is not a prediction, but simply a hypothetical scenario to illustrate why remedial CDR is so far from being interchangeable with mitigation now. Suppose that the atmospheric concentration of CO_2 continues to rise more than it would need to rise because of a short-sighted and self-centered policy choice now to gamble on less ambitious mitigation. The resultant "overshoot" in CO_2 is eventually reduced through remedial CDR. But during the temporary period of emissions overshoot, the additional atmospheric accumulation drives an increase in the temperature of the ocean water that bathes Totten Glacier at its grounding line sufficiently to precipitate the start of irreversible melting in the "cork" for a large chunk of East Antarctica as well.

Totten Glacier by itself contains roughly the same amount of water as the entire WAIS—equivalent to at least 3.5 m of global sea-level rise—so adding the melting of Totten to the melting of WAIS would double the amount of sea-level rise globally to 7 m (well over 20 feet, submerging tens of thousands of additional square miles of land into the oceans). That sea-level rise would endure for millennia—from a human perspective, forever. This despite the fact that the "overshoot" of atmospheric CO_2 that launched the melting was eventually "reversed" by remedial CDR. The temperature itself might even (much) later come back down, although only long after the accumulation of CO_2 was reduced, but the glacier might by then already have melted. Can we count on nothing like this hypothetical scenario happening? That is the gamble taken by self-indulgently unambitious mitigation now that relies on remedial CDR later.[54]

In the current context, what is particularly repugnant morally about this scenario is the following. Prior to any awareness of the apparent possibility of "recovery" from emissions "overshoot" through remedial CDR, certain generations might have decided to be less ambitious about mitigation simply out of preoccupation with themselves and a simple desire to avoid expense and disruption and to continue to enjoy convenience. We have seen that this would have amounted in fact to those generations taking a gamble in which only they themselves can gain and future generations can only lose. But those generations certainly might not have conceived of their actions in this way or been conscious that they were imposing such a bad gamble on future generations.

By contrast, choosing less ambitious mitigation now precisely because we are aware that it will allow the current generation to continue to produce avoidable emissions that we know will need to be reversed, only because we think future generations might be able to reverse them through remedial CDR, seems much worse, because more heartless and relentlessly self-preoccupied. Mitigating less now in full consciousness that, and precisely because, later generations might be able subsequently to reverse (some of) the additional emissions produced by the current less ambitious mitigation would be a clear case of choosing knowingly to make a problem worse, when one could easily have made it better, merely because the additional risks of severely worse climate change knowingly created will fall on other generations. Even if the subsequent reversal of the emissions overshoot itself by means of remedial CDR were entirely possible and completely acceptable, such inaction now would be the height of callous self-preoccupation and disregard of vulnerable others. If, instead, as illustrated by the East Antarctic scenario above, the overshoot's effects cannot be simply canceled out because in the interim physical tipping points are passed and more severe climate effects are ratcheted in, then such actions are outrageous. Why exactly outrageous?

Because worse than merely knowingly passing off the costs of one's own benefits to others (as in all externalization), one is in this temporal externalization knowingly passing off risks that are likely to compound and could pass the limits of social adaptation, as we saw late in chapter 1.

Alternatively, what an exciting privilege it is to belong to the pivotal generation with the opportunity to carry out a once-in-a-civilization transition, one of history's great revolutions, rivaling the Agricultural Revolution and the Industrial Revolution, whose dark side we would be confronting. The Energy Revolution, the building of a carbon-free energy regime, can potentially not merely limit the damage from climate change, but create a far cleaner, safer, healthier, smarter, more diverse, more technologically advanced, and perhaps even more just world. It will naturally have its own dark side—we are already finding, for instance, that the exotic materials in the batteries critical to renewables, like lithium and nickel, bring their own problems for both extraction and waste management. But if we can bequeath to future generations normal problems, not extraordinary dangers, we will have made an exceptional contribution. Members of future generations will be glad that we rose to the occasion and perhaps remember us proudly. We may, as Scheffler put it, have left the party early, but we will not have spoiled it for everyone else still to come.

Annette Baier once strikingly suggested that "morality is the culturally acquired art of selecting which harms to notice and worry about."[55] Some of the most difficult harms to notice, by far, are those we are locking in for future generations for as long as we continue to expand the cumulative atmospheric concentration of CO_2. The present *is* indeed the date-of-last-opportunity to set a limit on the extent of climate change by minimizing the emissions "overshoot" that could carry the planet beyond crucial points of no return that need not be passed if instead we move quickly enough to net zero carbon. The generations alive now therefore bear distinctively awesome responsibilities, but for the same reasons we also hold in our hands the power to launch a historic, transformative transition into a far less dangerous world than the fossil-fuel-fed nightmare that the energy-business-as-usual will otherwise yield. We now can create an invaluable legacy for the far future—and only we: later will be too late.

Radically different futures are possible, and the present generations will determine which versions of the future become more likely and which, less likely, by how passionately and wisely we act. Business-as-usual and toleration of the schemes of the dominant energy regime and its well-financed political enablers would produce a future of climate change that is increasingly dangerous—in fact, dangerous without limit—and likely feeding upon itself in abrupt and perhaps cascading departures from the weather and climate patterns that previously made what we perceive as normal possible for the geologically exceptional millennia of the now-ending Holocene.[56]

It is almost impossible not to underestimate the distance over which the reach of the present extends into the far future in the case of climate. We cannot see nearly as far as our influence stretches. We are willy-nilly making or allowing a world whose fundamental social structures and physical processes will endure until someone deliberately changes them, if they can. These basic structures and processes will shape which possibilities are and are not open, at least initially, to the people who follow us in multiple future generations.

Further change by them to reverse the effects of our action or inaction is not impossible in some cases, but it comes with various prices in money, effort, and—crucially—time lost between attempted initiation of the reversal and its culmination. Often modification in the inherited structures and processes can come only with substantial time lags, sometimes lags of more than a single generation—the generation that launches the modification does not live to see its outcome. And sometimes the effects of our choices are, from a practical, human perspective, irreversible. Nothing that future generations could possibly do would tame some forces that we may unleash, and only the remaining basic natural forces of the planet can—over geological time, during the passage of hundreds of human generations—slow or reverse some of our effects. William Blake exaggerated less than he may have seemed in saying, as the Industrial Revolution heated up, "The generations of men run on in the tide of Time / But leave their destin'd lineaments permanent for ever and ever."[57]

What, then, shall we do?[58]

5

Taking Control of Our Legacy

We should not expect the interest groups facing an existential threat from climate policy—fossil fuel companies and electric utilities—to go quietly into the night. . . . Opponents do not passively accept policy defeat. . . . Advocates cannot just focus on building the future. They must also dismantle the past. To transform our energy system and address the climate crisis head on, we must undermine opponents' political power.[1]

In battles over the shape of future energy systems the possibilities for democracy are at stake. . . . Such democratic struggles depend . . . upon identifying in current socio-technical systems their points of vulnerability.[2]

It is a serious error to over-estimate the intelligence and foresight, let alone the wisdom, of the rich and powerful.[3]

Powerful Enemies, Allied Masses

We face a fierce battle—not everyone is on the same side, by any means. The most unrelenting opponents of progress toward a net zero carbon world are fossil-fuel interests and their dedicated and entrenched allies in government and banking. We must no longer tolerate their deceptions, diversions, and detours. The struggle against them requires a broad mobilization of citizen energy: citizens joining and building social movements to replace the political and economic practices and structures that are blocking action on the climate. The political deck is now heavily stacked in favor of the energy status quo. The opponents of ambitious action to move quickly to net zero carbon are those with vested interests in fossil fuel, and they are among the wealthiest and most powerful individuals and organizations on earth. They are currently served by the heads of state of several critical nations—Morrison in Australia, Bolsonaro in Brazil, Putin in Russia, Mohammad bin Salman in Saudi Arabia, among others—and a considerable number of US senators and representatives. Bringing climate change under control will require tough political fights against ruthless, mendacious, and entrenched combinations of economic and political power.

We are examining general human responsibilities, and so far, I have commented explicitly on those of individuals and governments in wealthy nations like the United States whose wealth is heavily derived from industrial activities and from lifestyles that are driven by the combustion of fossil fuels and that, as a result, are leading past and present causes of climate change. Among these responsibilities are uncontroversial universal negative duties not to violate rights or otherwise harm others. The firms that market coal, oil, and gas are of course not exempt from the universal negative duties not to harm that apply to everyone else. In their case, the primary specific form that the duty takes is not marketing harmful products or, if it were somehow necessary to market some harmful products temporarily, to inform potential buyers of those products of the nature and extent of the harm that the use of the products would do.[4] Then the buyers could make an informed choice about whether to use the products, use them with appropriate restraint if they had to be used, and so forth.

The fossil-fuel companies have for decades flagrantly violated this universal minimal duty by knowingly pushing products that they understood long before most other people did are progressively undermining the stability of the climate, by systematically lying about how harmful the use of their products is, by viciously attacking scientists who have told the general public the truth, and by utterly failing to invest in measures that they also long ago understood well (like carbon capture and storage) that would have made their products safer.[5] So now that they have habitually demonstrated an incorrigible refusal to fulfill minimal duties, the rest of us have no choice but to force them to change their business plans or to go out of business. We now must defend our own health and safety and protect the basic rights of the people of future generations against the fossil-fuel interests in addition to fulfilling our own original responsibilities.

This is not the place for a systematic analysis of the political battlefield,[6] but a few quick examples may illuminate the situation. Reducing our own contributions to climate change through modifications in our individual consumption is definitely worthwhile,[7] but incremental, individual action is grossly insufficient even in aggregate. Bringing carbon emissions to net zero is not a matter merely of large numbers of individual consumers changing from one technology to another, like unscrewing incandescent bulbs and inserting LEDs, or even purchasing an electric or green hydrogen car. First, many more zero-carbon technologies need research, development, and demonstration that will often require supportive public policy. In order to make any significant difference, we need to build social movements to attack directly the economic and political structures that with all their influence and guile are blocking comprehensive political measures to reduce carbon emissions and to replace those impediments with public policies and institutions designed to protect the majority of people of future generations, not to enrich further a powerful minority in the present by protecting their business-as-usual.[8]

Unfortunately, while sales of fossil fuels may not be growing as rapidly as they used to, which is bad for their owners' stock prices, their use is not yet shrinking over the long term (outside the duration of the coronavirus pandemic), which is bad for the climate and most living beings. The use of renewable energy is growing rapidly, but in itself this does nothing whatsoever to slow climate change. If the renewable energy is consumed in addition to, not instead of, carbon energy, the carbon emissions undermining the climate continue unabated. And most renewable energy is in fact an addition, not a replacement: "Most policies tend to focus on supporting low-carbon alternatives, such as solar, wind, or electric vehicles, but these technologies often add to existing demand and therefore do not displace fossil-fuel use to any great extent. Public policies need to place far more importance on directly cutting back the use of fossil fuels or removing their emissions through CCS [carbon capture and storage], particularly the phasing out of coal power plants and conventional vehicles, well before they reach their productive end-of-life."[9]

Wallace-Wells puts it dramatically:

> The much-heralded green energy "revolution" . . . has yielded productivity gains in energy and cost reductions far beyond the predictions of even the most doe-eyed optimists, and yet has not even bent the curve of carbon emissions downward. We are, in other words, billions of dollars and thousands of dramatic breakthroughs later, precisely where we started when hippies were affixing solar panels to their geodesic domes. That is because the market has not responded to these developments by seamlessly retiring dirty energy sources and replacing them with clean ones. It has responded by simply adding the new capacity to the same system.[10]

This is bad news for everyone with naive faith that the use of fossil fuels would decline through some market magic if only renewables grew; and it is a clear example of the mistake, noted by Stokes in one of the epigraphs, of failing to dismantle the institutions from the past that otherwise block the way toward the institutions for the future. Current failing policies fit perfectly with the desire of the owners of fossil fuels to continue to extract and sell them and to avoid even belatedly investing seriously in CCS, while instead continuing, as always, to pour their resources into exploration for ever greater reserves to extract and sell in future.

In addition, major oil companies are vigorously resisting any decline in overall sales of their fossil fuels by moving increasingly into fracked gas for the manufacture of more plastics by their expanding petrochemical divisions.[11] Shell Polymers, for example, is constructing a multibillion-dollar ethane "cracker" plant on the Ohio River in Beaver County, Pennsylvania, to make ethylene for plastics from fracked gas.[12] The plastics manufacturers buy extensive ads in favor of recycling in order to create the false impression that the fundamental problem is not excess manufacturing and marketing of plastic, but negligent disposal by consumers, who have plastic virtually forced on them with everything they purchase.[13] And the carbon interests are aggressively pressuring African countries to accept both plastics and plastic waste that the countries wisely reject.[14] Fossil-fuel interests are also fighting back with other vicious tactics like campaigning state by state for harsh laws to block demonstrations against new pipelines by restricting freedom of speech and freedom of assembly and ignoring indigenous rights.[15]

Meanwhile, owners of some "zombie" pipelines and "orphaned" oil wells are engaging in a final act of destructive externalization extraordinaire. "Unlike most other industries, oil and gas is lawfully required to clean up its mess when the music stops," which means "required to plug the well in a manner that permanently confines all oil, gas and water in the separate strata in which they are originally found, and take other measures necessary to restore the location to a safe and clean condition."[16] Instead, executives of failed companies are paying themselves obscenely large sums, declaring bankruptcy, and walking away from their asset retirement obligations (AROs). For example, "Chesapeake Energy, which declared bankruptcy last month after paying out executive bonuses, might also be environmentally insolvent . . . with potential cleanup costs of $1.4 billion, nearly as much as its year-end market value of $1.6 billion. Chesapeake's filings show that it has set aside only $41 million in bonds to cover the cleanup of its 6,800 wells."[17] Performing the ARO for a typical fracked well costs around $300,000: "the true costs to plug deep U.S. shale wells may be one of industry's best kept secrets."[18] Taxpayers in the state in which the well is located will be left with the ARO costs—or with perpetually polluted land, groundwater, and air (from leaking methane).[19] These ARO-dodging executives thus bequeath to future generations "stranded liabilities" to complement their "stranded assets."[20]

Marketing carbon-based energy and carbon-based materials like plastics is the source of wealth for what is arguably the most powerful complex of multinational institutions in the world, with tentacles that reach deep into the main national and international political and economic structures. Many of the largest fossil-fuel companies are state owned, state controlled, or both: Saudi Arabia's Saudi Aramco, Russia's Gazprom and Rosneft, China's PetroChina and Sinopec, the National Iranian Oil Co., Norway's Equinor, and Coal India, for instance. "In 2018, nearly 60% of production of oil and 50% of production of gas came from state-controlled oil companies."[21] This crucial fact about state-ownership is often overlooked by climate activists.

Still, the "supermajors" among the investor-owned firms, namely Chevron, ExxonMobil, BP, Shell, and Total, are also huge. "The companies that supply and use fossil fuels make up a quarter of the global stock market and half the corporate bond market."[22] The US oil industry, for example, "directly and indirectly employs 10 million people."[23] ExxonMobil and Rosneft are deeply interlinked, leading ExxonMobil to support Vladimir Putin on crucial issues, as are other state-owned and investor-owned firms.[24] Then there is the privately owned but pervasively influential Koch Industries, one of the two largest private companies in the United States.[25]

The fossil-fuel companies themselves are buttressed by the service companies like Halliburton, Schlumberger, Transocean, and

Fluor; their insurers like Liberty Mutual; the hedge funds that buy the stocks, like BlackRock (the world's largest investor);[26] the other pipeline companies besides Koch; the other refineries besides Koch; the liquified natural gas facilities; the supertanker fleets; the freight lines with the coal cars and the oil tank-cars; the lobbyists (who glide through the revolving doors between regulator and regulated) and their giant PR firms; kept "think-tanks," kept "intellectuals," and front organizations, who tell the sweet lies; and crucially the banks that loan the capital for the industry's infrastructure and wait to be repaid over the infrastructure's long, useful life. It goes without saying that the state-owned companies are deeply embedded in the political and economic institutions of their respective powerful nations, but so are the investor-owned and privately owned companies on whose contributions many a US senator and representative depends to pay for campaign ads and social media operatives.[27]

For decades, public funds have gushed into the coffers of politically connected fossil-fuel companies as subsidies: global government subsidies in 2015 were $4.7 trillion, which was 6.3% of global economic product perversely wasted—unbelievably, more than the world's total expenditure on public health!—and the subsidies are projected to have been $5.2 trillion (6.5% of GDP) in 2017, and continuing.[28] According to the 2019 IMF study just cited, the US federal government handed over $649 billion from US taxpayers to fossil-fuel firms in 2015—this welfare system for carbon majors operated long prior to the Trump administration and its sustained assault on environmental, health, and safety regulations[29]—and the Russian federal government provided $551 billion in public money to the oil and gas companies favored by Putin. The conservative IMF calculates that "efficient fossil fuel pricing in 2015 would have lowered global carbon emissions by 28 percent and fossil fuel air pollution deaths by 46 percent."[30] The survival of these utterly inefficient and murderous subsidies are often the product of deep and inveterate political corruption.[31] The recession caused by the pandemic is initially wreaking havoc on fossil-fuel companies as it is on many others, but kept politicians can be counted on to bail the largest ("too big to fail") private and state companies out of their difficulties during the recession with more public funds, and without any conditions requiring compliance with even the minimal necessities for action against climate change.[32]

The necessary transformation of economic and political structures requires far more than for the market to accept noncarbon energy at its own admittedly quickening pace (in spite of the market-distorting subsidies and bailouts favoring fossil fuels). We must also, for example, get infrastructure for noncarbon energy production and transmission into developing nations who cannot afford both to adapt to the destructive climate changes that are already unleashed and to construct new energy infrastructure, so that their struggle to develop does not lock them into the infrastructure of the carbon energy regime that needs to be supplanted, as we saw in chapter 3. We need positive social action internationally, not only domestically.

Planning the "Final Harvest": Deceptive Delay

The fossil-fuel companies appear to have added new tactics for their fight against climate action. The fundamental strategy of the fossil-fuel interests always involves deceiving the general public. It is now well documented that corporate scientists understood the dangers of climate change decades before the public did, but company executives decided to sell as much fossil fuel as possible before the public caught on. The original strategy was to claim that the science was uncertain, as had been claimed earlier on behalf of tobacco companies about the science showing that smoking causes cancer, and to provide decades of funding to organizations that would minimize or deny the dangers and attack genuine climate scientists. An accessible and recent summary of decades of opposition by oil companies to action to protect the climate is *Smoke and Fumes: The Legal and Evidentiary Basis for Holding Big Oil Accountable for the Climate Crisis*.[33] Electric utilities, which had sunk their capital into coal-burning generation plants, also conducted a systematic campaign of deception.[34] And the sophisticated mendacity on behalf of oil and gas continues today.[35]

Meanwhile, newer tactics of strategic deception include pretending to be on board efforts to deal with climate change and claiming to be ready to help fulfill the Paris Agreement of 2015, but slowing down the transition to net zero carbon as much as possible in order, as before, to continue selling as much fossil fuel as possible—what Wallace-Wells calls "profiteering delay."[36] Deceptive delay has augmented deceptive denial. Once the world decarbonizes, the unsold fossil fuels that currently give the petrostates and the carbon majors their wealth may lose much of that value, becoming stranded assets. So it is in their own narrow interest to "move it before they lose it": "for Saudi Arabia, which has 75 years' worth of oil in the ground, the most rational approach may be to pump as hard as possible now."[37] Stokes has documented the general pattern: "these companies win by stalling."[38] Or as the Texas bumper sticker in the 1980s said, "Please, God, give us one more boom. We promise not to piss this one away."[39]

Three separate investigations reveal this fresh additional tactic in the misinformation strategy, instigated since the Paris Agreement of 2015, of pretending to cooperate in bringing down emissions while quietly carrying on exploration and extraction largely as usual and partly for newer purposes like plastics and "blue" hydrogen (hydrogen derived from natural gas, as distinguished from "green" hydrogen derived from water).

First, an investigation by the *Guardian* newspaper in 2019, prior to the recession caused by the coronavirus pandemic, concluded, "The world's 50 biggest oil companies are poised to flood markets with an additional 7m barrels a day over the next decade. . . . New research commissioned by the *Guardian* forecasts Shell and ExxonMobil will be among the leaders with a projected production increase of more than 35% between 2018 and 2030—a sharper rise than over the previous 12 years. The acceleration is almost the opposite of the 45% reduction in carbon emissions by 2030 that scientists say is necessary to have any chance of holding global heating at a relatively safe level of 1.5C. The projections are by Norwegian consultancy Rystad Energy, regarded as the gold standard for data in the industry."[40] University of Oxford energy economist Dieter Helm told the *Guardian*, "If we were serious about addressing climate change we would leave some oil in the ground, so there is a scramble among big oil companies to make sure their assets are not the ones left stranded."[41]

Second, complementary evidence is provided by government plans as distinguished from corporate plans as such, although of course the largest fossil-fuel corporations are, as we noted above, state enterprises. The latest edition of a comprehensive annual study led by the Stockholm Environment Institute (SEI) in cooperation with several other leading environmental research institutes and the United Nations Environment Program (UNEP), *The Production Gap 2020 Special Report*, conducted "extraction-based emissions accounting" that attributes emissions to the country within which the fossil fuel was extracted. Unusually, then, this study looks at the supply side of the fossil-fuel regime. The key finding of the study is "to follow a 1.5°C-consistent pathway, the world will need to decrease fossil fuel production by roughly 6% per year between 2020 and 2030. Countries are instead planning and projecting an average annual *increase* of 2%, which by 2030 would result in more than double the production consistent with the 1.5°C limit."[42] Although subsidies for fossil-fuel consumption have recently been declining, "fossil fuel production subsidies are actually on the rise. . . . Leading providers of fossil fuel subsidies, by quantified monetary value, include Canada, China, Russia, and the United States."[43]

The need for government economic support for recovery from the economic recession caused by COVID-19 presented a unique historic opportunity, as the slogan has it, "to build back better" by creating jobs in alternative energy sectors rather than bailing out fossil-fuel enterprises that need to be eliminated fairly soon and that, in many cases, were already struggling before COVID-19 struck. At the global level, this rare opportunity is, however, being thrown away: "Despite widespread calls for a green recovery, as of November 2020, governments have directed more support to fossil fuel and other carbon-intensive activities than to clean energy and low-carbon sectors. One recent assessment suggests that most countries' stimulus packages have been 'environmentally harmful' overall. . . . Another assessment finds that, as of early November 2020, government stimulus and recovery packages had committed nearly five times more to high-carbon sectors, such as fossil fuel extraction, aviation, and car manufacturing (USD 878 billion), than to low-carbon industries, such as electric vehicles, energy efficiency, and renewable energy (USD 179 billion)."[44] This is a tragic and monumental failure in public policy that can be corrected most effectively by defeating incumbent politicians.

Third, the clearest evidence of all about the fossil-fuel companies' strategy is of course their behavior. Many oil and gas companies trumpet in their social media and television advertising that they are rapidly diversifying and turning themselves into "energy companies" rather than fossil-fuel companies by investing in renewables.[45] Instead, a recent analysis by the staid International Energy Agency (IEA) found this: "For the group of companies analysed, aggregate annual capital expenditures for projects outside core oil and gas supply averaged under USD 2 billion since 2015, less than 1% of the total capital expenditures by these companies, though some companies have spent up to an average of 5%, and the total topped USD 2 billion for the first time in 2019."[46]

Put the other way around, in the most recent years since the *Paris Agreement* in 2015, leading oil and gas companies have devoted an average of 99% of their capital expenditures into finding and selling more oil and gas, although some managed to restrict their expansion into carbon fuel to a mere 95% of their total investment![47] Where they put their money is decisive evidence of which side of the struggle to eliminate carbon emissions they are on. Meanwhile, they assure us through their massive public relations operations that they are serious about cutting carbon emissions, that they have adopted a goal of net zero carbon, and that they are changing their fundamental nature. If their actual behavior qualified as gradualism, it would be the most leisurely gradualism imaginable and totally inappropriate to our stringent circumstances. It is instead obstructionism and dissembling, designed once again to deceive us into passivity while the oil and gas continue to flow alongside the sun and wind.

Enabling the "Final Harvest": Banks' BAU

If one continues to follow the money, capital investment naturally leads back to the major holders of capital: banks, hedge funds, and other financial institutions who loan the fossil-fuel regime most of the funds that enable capital investment in greater production. On sincerity, the banks and hedge funds score a little higher than the fossil-fuel companies because they mostly have not even pretended to care about climate change and have simply continued business-as usual (BAU). A few banks have recently pledged not to loan to companies planning to drill for oil in the Arctic National Wildlife Refuge,[48] although this hardly counts as even a serious gesture, because drilling in a wildlife refuge when excess production has driven oil prices to their lowest level ever would be sheer vandalism.

On the whole, banks, led by US banks, continue to pour money into additional exploration, production, and transport of oil and gas. *Banking on Climate Change*, an annual report on research led by the Rainforest Action Network, found that "35 banks from Canada, China, Europe, Japan, and the U.S. have together funnelled USD $2.7 trillion into fossil fuels in the four years since the Paris Agreement was adopted (2016–2019)."[49] This is flagrantly irresponsible. The worst enabler of additional fossil fuel by far is JP Morgan Chase, which has loaned $269 billion just since the Paris Agreement in 2015, followed, in order, by Wells Fargo, Citi, Bank of America, RBC (Royal Bank of Canada), MUFG (Japan), and Barclays (Europe's most flagrant enabler).[50] The largest increases in funding in 2019 were by Bank of America and BNP Paribas.[51] The big four Chinese banks—Agricultural Bank of China, Bank of China, China Construction Bank, and ICBC (Industrial and Commercial Bank of China, the largest bank in the world)—gave the most support to the most primitive and dirtiest fossil fuel, coal, financing more than half of the world's coal mining and coal power.[52]

The three largest recipients of fossil-fuel loans were Occidental Petroleum (expansion of fracking), Enbridge (Line 3 tar sands oil pipeline in Minnesota), and TC Energy (Keystone XL tar sands oil pipeline), all for projects that ought to be stopped.[53] Bank loans for capital expenditure on new pipelines are especially destructive of the climate because of their long time frames. The typical loan is for decades, and it can be repaid only if the infrastructure that it makes possible is used throughout the decades of the loan's lifetime. Obviously, this relies on decades of additional carbon emissions when carbon emissions ought instead to be rapidly approaching net zero. Here again, present action reaches deep into the future to cause damage. Further, what economists would call the "opportunity costs" of current bank policies are monstrous: all these funds could be underwriting alternative energy, which is already competitive in price nearly everywhere and only needs more investment and mildly supportive government policies.

Exactly how to defend future generations against the current destructive bank practices is still being investigated and experimented with, but social movements are taking shape.[54] Meanwhile, we should minimally demand that all institutions in which we are involved, including local governments, withdraw all funds from the most grievously offending banks, move the funds to local banks and credit unions that do not loan to fossil-fuel companies, and refuse to apply for mortgages, loans, or credit cards from the banks underwriting the worst environmental destruction. Divestment stigmatized fossil-fuel companies, so perhaps highly visible boycotts can help to stigmatize irresponsible banks.[55] The vulnerability of future generations is so extreme that we cannot in good conscience continue to do so little to protect them, and the time for decisive action is so short that we cannot tolerate leisurely gradualism, much less deception, on the part of the carbon majors and myopically obstinate greed on the part of the state and multinational banks.

Preventing the "Final Harvest":
The Active Many vs. the Ruthless Few

Efforts to control climate change, then, face entrenched and interlinked opposition by major incumbent governments and major fossil-fuel firms and their financial supporters. I have focused here on why they must be confronted directly. This global web of vested, interlocked interests in wealth and power is working against those of us trying to limit climate change. It will take a while to extricate our politics and institutions from the grip of this strong and slippery politico-economic web. The fossil-fuel firms are profoundly recalcitrant institutions whose core business is marketing products that produce carbon emissions and whose very existence is threatened by action to limit climate change soon, if they continue to persist in refusing to make radical changes to their business plans. That overcoming opponents this well-entrenched, determined, and ruthless, who have systematically lied to the public for decades, will take some time is one more reason why the present is the date-of-last-opportunity to launch vigorous and determined exposure and opposition. Time is in short supply, so the fight needs to be intensified now.

The grubby, greedy, grasping, dog-eat-dog version of extractive capitalism practiced by firms in the fossil-fuel industry like ExxonMobil, Shell, Koch Industries, Gazprom, and Saudi Aramco are an embarrassment to decent business people. In order to resist them, one need not be anticapitalist—only antirapacity.[56] What are undoubtedly needed are broad structural changes to the energy system. Because of the wealth and power of the carbon status quo, many of the changes needed require government action, which presupposes a government with a level of responsiveness to its citizens that does not currently exist in Australia, Brazil, Russia, Saudi Arabia, the US Senate and House, and elsewhere. The most important action that readers of this book, who are likely to live in one of the countries with some degree of democracy, can take is to join with others to remove from political power as many as possible of the friends of fossil fuel. Matto Mildenberger has put this crisply: "We need to *first* disrupt the political power of carbon polluters before we can effectively reshape economic incentive structures. . . . Carbon polluters will not voluntarily relinquish their power. . . . Climate change mitigation is a political act. . . . Effective policies must reduce carbon pollution while also disrupting opponents' political power."[57] Fossil fuels cannot compete on their own: more than ever, they need the political favoritism and subsidies that they have long enjoyed. Political measures have lifted them up, and political measures can bring them down.

Every election now needs to be about climate change. Every candidate who refuses to commit to ambitious mitigation of carbon emissions and to plans for rapid progress toward net zero carbon must be soundly defeated and replaced with someone who is strongly committed to limiting damage to the climate. This means running for office, encouraging other good people to run for office, contributing to the financing of campaigns, campaigning, and fully exposing in all available media the rampant favoritism and corruption. For instance, in order to expose the corrupt relationships between the supermajors and vicious dictatorships like the Obiangs in Equatorial Guinea, Section 1504 of the 2010 Dodd-Frank Wall Street reform law, originally sponsored by Republican Senator Richard Lugar, needs to be restored. The elimination of this requirement of transparency was one of the first and most valuable gifts in 2017 by Congress and President Trump to Big Oil.[58] Social and political change, not just individual behavioral change, are essential.

Nevertheless, according to some highly respected financial analysts, the fossil-fuel regime is beginning to stumble for partly economic reasons. If so, we still need to give it several very hard shoves to make sure it goes all the way down! Such sector disruption often comes sooner than one might expect and occurs at or shortly before peak demand. "The fossil fuel system is ripe for disruption. It is low growth, high fixed cost, low return and (incredibly) planning on expansion even as demand peaks. The entire system is being disrupted by the forces of cheaper renewable technologies and more aggressive government policies. In one sector after another these are driving peak demand, which leads to lower prices, less profit, and stranded assets."[59] "The cheapest source of bulk electricity production in 85% of the world is renewables. . . . Nearly three quarters of new global electricity capacity is now non-fossil. This is important because electricity is the largest share of fossil fuel consumption, accounting for over one third of the total."[60] Solar costs have dropped 99% in the last four decades.[61]

Institutional investors in the stock market look for companies that can expect long-term growth in demand for their products. Fossil-fuel companies are approaching, if they have not already reached, peak demand because renewables are soaking up most of the growth in demand. "The energy sector has been a weak performer in the S & P 500 for a decade [2010–2019], and the supermajors' [Chevron, ExxonMobil, BP, Total, and Royal Dutch Shell] lackluster results have contributed to that underperformance."[62] ExxonMobil's "market value today is about a third of what it was in 2008, when it approached $500 billion,"[63] and its stock has been humiliatingly dropped from the calculation of the Dow Jones Industrial Average.[64]

The heartless strategy of attempting to prop up demand for fracked gas by flooding the planet, especially relatively defenseless African nations, with vastly more plastic is considered by hard-headed analysts at the Carbon Tracker Initiative to be likely to fail, leaving the oil and gas companies with still more stranded assets, this time in the form of excessive capacity in the new petrochemical installations now being hastily constructed (relying, of course, on bank loans—ones that appear unlikely ever to be repaid). Only about 5–10% of plastics are actually recycled, which is one more huge negative externality inflicted on society by fossil fuels; and manufacturing 1 ton of plastic produces approximately 5 tons of CO_2, which would mean that on current business plans, the plastics industry would use 19% of the total remaining cumulative carbon budget by 2040, which would be an outrageously wasteful misuse of a shrinking asset.[65] The Carbon Tracker Analyst Note just cited calmly concludes, "To have one sector planning on doubling its carbon footprint while the rest of the world plans to phase out emissions clearly makes no sense"[66]; and "this [plastics] is not an industry which has focussed at all on efficiency or maximising utility. It is a bloated behemoth, ripe for disruption."[67]

The Big Banks need to be persuaded to stop denying the reality of how risky long-term loans to fossil-fuel companies have become. Restoration of Section 1504 would contribute by revealing the true basis of some extraction contracts with poor countries in bribery and other forms of corruption. Research done for Finance Watch has recommended that banks increase their risk weightings for coal, oil, and gas exposure in order to avoid what it calls the "climate-finance doom loop," in which "fossil fuel finance enables climate change, and climate change threatens financial stability, through disruptive natural events. . . . Banks would therefore have more protection against the risk of carbon assets becoming 'stranded' if demand falls—or the risk of costly climate disruption if it does not."[68] So one message for the bankers who have not comprehended what is happening is that even if they do not care about the rest of us or the planet itself, they might want to stop shooting themselves in the foot with increasingly risky long-term loans for unsustainable projects, including both exploration/extraction and rushed construction of petrochemical plants to manufacture ocean-clogging plastics. That climate change is a threat to the fundamental stability of financial markets is being recognized increasingly widely, even in some cases by establishment institutions directed by Trump appointees, such as the Commodity Futures Trading Commission.[69]

We cannot have an all-purpose tactical program. We must be agile, innovative, and quick to pounce into openings. Leif Wenar, writing about a different but related problem, provides beautiful examples of tactical imaginativeness, such as what he calls "tapering."[70] Naomi Klein emphasizes analyzing the system for "chokepoints,"[71] and Timothy Mitchell brilliantly shows how the "points of vulnerability," as he calls them in one of our epigraphs, available to the British coal miners during the Industrial Revolution when coal was absolutely essential enabled them to use general strikes to enhance democracy generally as well as to improve conditions for themselves.[72]

And there are many more of us who are threatened than there are of them who are profiteering. As Jonathan Schell put it in another context, "The active many can overcome the ruthless few."[73] Once political and financial decisions begin to be made more nearly on their merits, dirty and damaging fossil fuels can be driven into decline.[74] Extraction and import can be regulated, for example, to require the removal from the atmosphere by the seller at its own expense of amounts of CO_2 at least equal to the CO_2 content of any fuel sold, finally limiting additional harm.[75] In the United States, once political dominance at the federal level has been wrestled away from the friends of fossil fuel, outrageous subsidies can be ended, and pollution controls arbitrarily and capriciously gutted by the Trump Administration can be restored, strengthened, and enforced. The United States can show leadership under the Paris Agreement and greatly strengthen the weak NDCs made by the Obama Administration. And much, much more—vastly increased requirements for transparency about plans to reach net zero enforced by financial regulators, lawsuits on a wide variety of grounds, increasing divestment, boycotts, demonstrations, and relentless exposure of actual behavior to resist relentless deception.[76]

I was originally drafting this paragraph on Earth Day 2020. My last employer, the University of Oxford, announced that it would belatedly divest completely from fossil fuels. And, showing some genuine leadership, Oxford also instructed its endowment fund managers to invest across the board only in corporations with net zero business plans. (We are still waiting for Harvard.) To some extent, this is only symbolic, but it does demonstrate that even ponderous institutions can be persuaded and pressured into changing. Net zero business plans will work better when governments are seriously committed to net zero public policies (as the UK government claims to be), and vice versa. Removing from power all the politicians who are blocking serious action to reduce emissions rapidly and forcing carbon majors to change radically or go out of business is still an uphill battle. But governments are supposed to protect their people, so it is nevertheless a battle that must be won, and soon, especially for the sake of vulnerable future people whose fates are held in the hands of the living. We can perhaps still act just in time.

It is very important that it never becomes likely that the earth's climate will run wildly out of control. We do not know exactly how urgent action is because of the very uncertainties that some opponents of action irrationally invoke in support of business-as-usual. The time to establish a limit on climate change is now—while we still can. The stakes are far too high relative to the insignificance of the lifestyle we need to give up in order to return human civilization to a much safer place. Pogo—I show my age again—said, "We have met the enemy and he is us."[77] This is profoundly true of climate change because our passivity and inattention have allowed fossil-fuel interests to dominate energy policy and energy politics for a century. But it could also become true that we have met the allies and they are us. We have agency—our response to our time is our choice. The direction the future takes is up to us, if our pivotal generation takes back the initiative from the entrenched interests who will undermine the climate rather than willingly surrender any of their wealth and power.

A clear view of the situation raises the practical questions: who is going to lead the turn away from the path that may have a maelstrom at its unseen end, and when are they going to get started?

The answer, I have tried to show, ought to be: us now. We are the pivotal generation, and the responsibility falls on us. We can recapture control of our destiny and our legacy by restoring democratic control of our politics and accelerating the revolutionary energy transition that will brighten the human future. This is the crucial political fight of the twenty-first century. It is too important to lose from lack of thought, effort, and endurance. We can do this, but we have to start doing it now.

ACKNOWLEDGMENTS

Chapter 1 includes greatly revised versions of arguments originally presented in the 2017 John Dewey Lecture at the Law School of the University of Chicago and in a 2017 James A. Moffett '29 Lecture in Ethics at Princeton University. Neither lecture has been published. It also develops themes about time initially introduced in Henry Shue, "Responsibility to Future Generations and the Technological Transition," in *Perspectives on Climate Change: Science, Economics, Politics, Ethics*, ed. Walter Sinnott-Armstrong and Richard B. Howarth (Amsterdam and San Diego: Elsevier, 2005), 265–283; reprinted in Henry Shue, *Climate Justice: Vulnerability and Protection* (Oxford: Oxford University Press, 2014), 225–243.

Chapter 2 draws upon arguments originally introduced in Henry Shue, "Historical Responsibility," Technical Briefing for Ad Hoc Working Group on Long-term Cooperative Action under the Convention (AWG-LCA), SBSTA, UNFCC, Bonn, 4 June 2009, http://unfccc.int/files/meetings/ad_hoc_working_groups/lca /application/pdf/1_shue_rev.pdf; and it extensively reworks and elaborates the version of those arguments formulated in Henry Shue, "Historical Responsibility, Harm Prohibition, and Preservation Requirement: Core Practical Convergence on Climate Change," *Moral Philosophy and Politics* 2 (2015), 7–31, doi:10.1515/ mop-2013–0009. Two paragraphs in chapter 1 also take sentences from the latter.

Chapter 3 draws upon, substantially revises, and expands arguments from Henry Shue, "Distant Strangers and the Illusion of Separation: Climate, Development, and Disaster," in *The Oxford*

Handbook of Global Justice, ed. Thom Brooks (Oxford: Oxford University Press, 2020), 259–276.

Chapter 4 draws upon, substantially reformulates, and integrates arguments from two earlier articles: Henry Shue, "Climate Dreaming: Negative Emissions, Risk Transfer, and Irreversibility," *Journal of Human Rights and Environment* 8 (2017), 203–216, doi:10.4337/jhre.2017.02.02; and Henry Shue, "Mitigation Gambles: Uncertainty, Urgency, and the Last Gamble Possible," *Philosophical Transactions of the Royal Society A* 376 (2018), 1–11, doi:10.1098/rsta.2017.0105.

APPENDIX ON INEQUALITY

Chapter 2 discusses the extent of inequality among nations in causing climate change, which is normally measured in cumulative national carbon emissions, or cumulative national per capita carbon emissions, and which has resulted primarily from the fact that some nations industrialized earlier and became much wealthier than others. Because climate negotiations have mostly been conducted by national governments in international fora, these inequalities in historical national responsibility have received some attention, if insufficient action.

Inequalities in wealth among individuals, however, are even more outrageously extreme than inequalities in wealth among nations. Some very wealthy individuals live in very poor nations— indeed, frequently the nations remain poor in part because the families of their dictators systematically steal the national wealth by, for example, diverting the proceeds of oil drilling, mining, and other extraction of natural resources.[1] What may be a little less widely understood is that differences in individual wealth lead by way of differences in consumption choices to differences in the extent of causation of climate change.

Research reported in 2020 by the Stockholm Environment Institute and Oxfam has led to some startling estimates.[2] The findings for the years 1990–2015 included the following:

> The richest 10% of the world's population (c. 630 million people) were responsible for 52% of the cumulative carbon emissions—depleting the global carbon budget by nearly a third (31%) in those 25 years alone;

The poorest 50% (c. 3.1 billion people) were responsible for just 7% of cumulative emissions, and used just 4% of the available carbon budget;

The richest 1% (c. 63 million people) alone were responsible for 15% of cumulative emissions, and 9% of the carbon budget—twice as much as the poorest half of the world's population;

The richest 5% (c. 315 million people) were responsible for over a third (37%) of the total growth in emissions, while the total growth in emissions of the richest 1% was three times that of the poorest 50%.[3]

Oxfam and SEI conclude,

The global carbon budget is being rapidly depleted, not for the purpose of lifting all of humanity to a decent standard of living, but to a large extent to expand the consumption of a minority of the world's very richest people. . . . Doubling the per capita [carbon] footprint of the poorest 50% of the world's population from 1990 to 2015 would have increased total global emissions by *less than* the growth in emissions associated with the richest 1% in this period.[4]

These differences in the consumption of the rich and the poor mean that some measures to reduce carbon emissions can rightly be targeted specifically at what I have called the "luxury emissions" of the wealthy rather than the "subsistence emissions" of the poor.[5] Among the measures endorsed by Oxfam and SEI are "wealth taxes, luxury carbon taxes—such as carbon sales taxes on SUVs, private jets or super yachts, or levies on business class or frequent flights."[6]

A careful economic explanation for the kind of findings made by Oxfam and SEI is laid out in another recent study, whose basic conclusion is that "inequality in the distribution of energy footprints varies across different goods and services. Energy-intensive goods tend to be more elastic, leading to higher energy footprints of high-income individuals. . . . People with different purchasing power make use of different goods and services, which are sustained by different energy quantities and carriers."[7] A recent thought-piece offers an accessible account of the economic assumptions and sketches a number of imaginative rationales that converge on various luxury carbon taxes.[8]

As I have emphasized throughout, we must before very long reach net zero carbon emissions with no one emitting any CO_2 that is avoidable. But we have strong economic and ethical reasons for moving immediately against the superfluous luxury emissions of the affluent as one part of a broad program. Self-indulgent unnecessary emissions by the wealthy few are morally intolerable when they are threatening the climate for everyone from now on. Momentary selfishness causes millennial damage. We are responsible for protecting our defenseless descendants against our careless contemporaries. If not us, who? If not now, when?

NOTES

Chapter 1: The Pivotal Generation

1. Annette Baier, "The Rights of Past and Future Persons," in *Responsibilities to Future Generations: Environmental Ethics*, ed. Ernest Partridge (Buffalo, NY: Prometheus Books, 1981), 171–183, at 177.

2. Andreas Malm, *The Progress of This Storm: Nature and Society in a Warming World* (London: Verso, 2018), 1.

3. Elizabeth Kolbert, "What Will Another Decade of Climate Crisis Bring?," *New Yorker*, 5 January 2020, https://www.newyorker.com/magazine/2020/01/13/what-will-another-decade-of-climate-crisis-bring.

4. Jonathan Watts, Jillian Ambrose, and Adam Vaughan, "Oil Companies Scrambling to Raise Output in Final 'Fossil Fuel Harvest,'" *Guardian*, 11 October 2019, Special Investigation: The Polluters, 12–13. The significance of the plan for a final harvest of fossil fuel is discussed in chapter 5.

5. David Hume, *A Treatise of Human Nature*, ed. L. A. Selby-Bigge (Oxford: Oxford University Press, 1968), 437.

6. In focusing specifically on responsibility, I am restricting myself to only one of the four elements in Simon Caney's fourfold schema for a normative account of climate change; see his "Climate Change and Non-Ideal Theory: Six Ways of Responding to Non-Compliance," in *Climate Justice in a Non-Ideal World*, ed. Clare Heyward and Dominic Roser (Oxford: Oxford University Press, 2016), 21–42.

7. William Faulkner, *Requiem for a Nun* (London: Vintage, 1996), act 1, scene 3, 85.

8. Timothy Mitchell, "Afterword to the Paperback Edition," in *Carbon Democracy: Political Power in the Age of Oil* (New York: Verso, 2013), 260.

9. William Shakespeare, *Julius Caesar*, *The Complete Plays and Poems of William Shakespeare*, ed. William Allan Neilson and Charles Jarvis Hill (Cambridge MA: Riverside Press, 1942), act 3, scene 2, ll. 80–81.

10. William Shakespeare, "Sonnet 73," *Complete Plays and Poems of William Shakespeare*.

11. See, for example, Friederike Otto with Benjamin von Brackel, *Angry Weather: Heat Waves, Floods, Storms, and the New Science of Climate Change*, trans. Sarah Pybus (Berkeley and Vancouver: Greystone Books, 2020).

12. For two wonderfully different ways of showing this, see David Wallace-Wells, *The Uninhabitable Earth: A Story of the Future* (New York: Tim Duggan Books, 2019);

and Mary Robinson with Caitríona Palmer, *Climate Justice: Hope, Resilience and the Fight for a Sustainable Future* (New York: Bloomsbury, 2018).

13. BP, *Statistical Review of World Energy*, 69th ed. (2020), "All Data 1965–2019, Carbon Dioxide Emissions," https://www.bp.com/en/global/corporate/energy-economics/statistical-review-of-world-energy.html. Click on download titled "Statistical Review of World Energy—All Data, 1965–2019"; choose "Carbon Dioxide Emissions (from 1965)"; inspect the line labeled "Total World" and note that total for 2019 (34,169) is the largest total since 1965.

14. See "Trends in Atmospheric Carbon Dioxide," NOAA, Earth System Research Laboratories, Global Monitoring Laboratory, 5 April 2021, https://www.esrl.noaa.gov/gmd/ccgg/trends/monthly.html. Also see Henry Fountain, " 'Like Trash in a Landfill': Carbon Dioxide Keeps Piling Up in the Atmosphere," *New York Times*, 4 June 2020; and "Global Carbon Budget: Data," Global Carbon Project, 11 December 2020, https://www.globalcarbonproject.org/carbonbudget/20/data.htm.

15. "NOAA Annual Greenhouse Gas Index [AGGI]," NOAA, Earth System Research Laboratories, 2020, table 2, "Global Radiative Forcing, CO_2 Equivalent Mixing Ratio, and the AGGI 1979–2019," Spring 2020, https://www.esrl.noaa.gov/gmd/aggi/aggi.html. Also see David King and Rick Parnell, "Stopping Climate Change Could Cost Less Than Fighting Covid-19," *Washington Post*, 17 September 2020.

16. Myles R. Allen et al., "Framing and Context," in *Global Warming of 1.5 °C*, ed. Valérie Masson-Delmotte et al., Special Report (Intergovernmental Panel on Climate Change, 2018), 49–91, https://www.ipcc.ch/sr15/chapter/chapter-1/.

17. Myles Allen et al., "Summary for Policymakers," in Masson-Delmotte, *Global Warming of 1.5 °C*, 1–24, https://www.ipcc.ch/sr15/chapter/spm/.

18. For links to a collection of relevant earlier articles, see "The Climate Connection to California's Wildfires," *New York Times*, 8 September 2020.

19. A prescient discussion of these issues is Catriona McKinnon, *Climate Change and Future Justice: Precaution, Compensation, and Triage* (Abingdon: Routledge, 2012).

20. B. Ekwurzel et al., "The Rise in Global Atmospheric CO_2, Surface Temperature and Sea Level from Emissions Traced to Major Carbon Producers," *Climatic Change* 144 (2017), 579–590, at 587, doi:10.1007/s10584-017-1978-0.

21. Chapter 4 is devoted to CDR, which the integrated assessment models (IAMs) used by the Intergovernmental Panel on Climate Change (IPCC) invoke most heavily. Each of the other technologies raises its own specific issues, which I do not have space here to examine.

22. Martin Luther King Jr., "I Have a Dream," March on Washington, 28 August 1963, https://www.americanrhetoric.com/speeches/mlkihaveadream.htm.

23. John Rawls, *A Theory of Justice* (Cambridge, MA: Belknap Press of Harvard University Press, 1971), 289: "They try to piece together a just savings schedule by balancing how much at each stage they would be willing to save for their immediate descendants against what they would feel entitled to claim of their immediate predecessors."

24. Whoever the individuals turn out to be—Rawls correctly takes for granted that basic rights do not depend on individual identities.

25. Rawls, *Theory of Justice*, 287.

26. It may be worth highlighting at an abstract level the difference between tasks that cannot be passed on to future generations, which are the subject of the first observation about Rawls, and burdens and risks that can be—and sometimes inevitably are—passed on to future generations, which are the subject of the third observation. What I refer to as the "integral fabric" of history means that a challenge sometimes arises in one particular generation, and if the challenge is not dealt with then, it never can be dealt with because a unique opportunity is lost and circumstances move on. In chapter 3, I will introduce the notion of the date-of-last-opportunity to focus on such last chances. For example, if nation A is invaded by nation B, the current generation of nation A confronts the task of defending against the invasion. They cannot pass on to the next generation the task of defense against the invasion because their failure to defend would mean that nation A is conquered, and the opportunity for successful defense is lost. The next generation of nation A may confront the task of rebelling against entrenched conquerors and expelling them, but that is a different and presumably more difficult task.

My first observation, then, is that some *tasks* are not transferable across time: the defense must occur now before the conquest is completed. The third observation is that some *burdens* are transferable, and, most notably, the failure to perform a nontransferable *task* (like defense) transfers to some future generation the *burden* of providing for the nation's autonomy (by expelling entrenched conquerors). It makes no sense—except perhaps as a sort of curse against the universe—to complain that facing a nontransferable task is unfair, because this kind of task cannot be redistributed. But it can make complete sense to complain that bearing a transferable burden is unfair if an earlier generation was responsible for dealing with it and failed to do so, thereby dumping onto you the burden, which may now require performance of a more difficult task than the task that those originally responsible failed to perform. I am indebted to Catriona McKinnon for spotting the possible confusion here.

27. Roger Revelle and Hans E. Suess, "Carbon Dioxide Exchange between Atmosphere and Ocean and the Question of an Increase of Atmospheric CO_2 during the Past Decades," *Tellus* 9 (1957), 18–27, at 19.

28. William K. Stevens, "Scientist at Work: Wallace S. Broecker; Iconoclastic Guru of the Climate Debate," *New York Times*, 17 March 1998.

29. See Paul J. Crutzen, "Geology of Mankind," *Nature* 415 (2002), 23: "It seems appropriate to assign the term 'Anthropocene' to the present, in many ways human-dominated, geological epoch. . . . Unless there is a global catastrophe . . . mankind will remain a major environmental force for many millennia. . . . At this stage, however, we are still largely treading on terra incognita."

30. Joeri Rogelj et al., "Energy System Transformations for Limiting End-of-Century Warming to below 1.5 °C," *Nature Climate Change* 5 (2015; corrected 2016), 519–528 and 538 (corrigendum), doi:10.1038/nclimate2572; and Joeri Rogelj et al., "Differences between Carbon Budget Estimates Unravelled," *Nature Climate Change* 6 (2016), 245–252, doi:10.1038/nclimate2868.

31. For wonderfully lucid and imaginative explanations of planetary mechanics, see Richard B. Alley, *The Two-Mile Time Machine: Ice Cores, Abrupt Climate Change, and Our Future*, rev. ed. (Princeton, NJ: Princeton University Press, 2014).

32. Thomas Hobbes, *Leviathan or the Matter, Forme and Power of a Commonwealth Ecclesiasticall and Civil* (Oxford: Basil Blackwell, n.d. [1651]), First Part, Chapter XIII, 82.

33. Geoffrey Parker, *Global Crisis: War, Climate Change & Catastrophe in the Seventeenth Century* (New Haven, CT: Yale University Press, 2014), xxvi and 28–29; and Geoffrey Parker, *Global Crisis: War, Climate Change and Catastrophe in the Seventeenth Century*, abr. and rev. ed. (New Haven, CT: Yale University Press, 2017), xxi and 26. Also see Sam White, *A Cold Welcome: The Little Ice Age and Europe's Encounter with North America* (Cambridge, MA: Harvard University Press, 2017); and Philipp Blom, *Nature's Mutiny: How the Little Ice Age Transformed the West and Shaped the Present* (New York: W. W. Norton, 2019).

34. Michael T. Klare, *All Hell Breaking Loose: The Pentagon's Perspective on Climate Change* (New York: Metropolitan Books, 2019), 40–48.

35. Timothy Snyder, *Black Earth: The Holocaust as History and Warning* (New York: Vintage Books, 2016), 325–326.

36. For powerful considerations in favor of taking possible extinction very seriously, see Toby Ord, *The Precipice: Existential Risk and the Future of Humanity* (New York: Bloomsbury Publishing, 2020). One exploration of unraveling is Geoff Mann and Joel Wainwright, *Climate Leviathan: A Political Theory of Our Planetary Future* (London: Verso, 2018).

37. See Ayana Elizabeth Johnson and Katharine Keeble Wilkinson, eds., *All We Can Save: Truth, Courage, and Solutions to the Climate Crisis* (New York: One World, 2020).

38. Henry Shue, "Deadly Delays, Saving Opportunities: Creating a More Dangerous World?" (2010), reprinted in *Climate Justice: Vulnerability and Protection* (Oxford: Oxford University Press, 2014), 263–286.

39. Lauren Hartzell-Nichols, *A Climate of Risk: Precautionary Principles, Catastrophes, and Climate Change* (New York: Routledge, 2017).

40. For saving me from a serious error at this point, I am immensely grateful to Catriona McKinnon.

41. Simon Caney, "Justice and Posterity," in *Climate Justice: Integrating Economics and Philosophy*, ed. Ravi Kanbur and Henry Shue (Oxford: Oxford University Press, 2019), 157–174.

42. See Timothy M. Lenton et al., "Climate Tipping Points—Too Risky to Bet against," *Nature* 575 (2019; corrected 2020), 592–595; and further references there. Also see Zeke Hausfather and Richard Betts, "Analysis: How 'Carbon-Cycle Feedbacks' Could Make Global Warming Worse," Carbon Brief, 14 April 2020, https://www.carbonbrief.org/analysis-how-carbon-cycle-feedbacks-could-make-global-warming-worse. Hausfather and Betts note, "Analysis for this article shows that feedbacks could result in up to 25% more warming than in the main IPCC projections."

43. Malm, *Progress of This Storm*, 73.

44. Thus the two requirements specified in Shue, "Deadly Delays, Saving Opportunities," are satisfied. The classic study is Timothy M. Lenton et al., "Tipping Elements

in the Earth's Climate System," *Proceedings of the National Academy of Sciences of the USA* 105 (2008), 1786–1793. Also see David Frame and Myles R. Allen, "Climate Change and Global Risk," in *Global Catastrophic Risk*, ed. Nick Bostrom and Milan M. Ćirković (Oxford: Oxford University Press, 2008), 273–275. On tipping points for Antarctica and Greenland, see Frank Pattyn et al., "The Greenland and Antarctic Ice Sheets under 1.5 °C Global Warming: Review Article," *Nature Climate Change* 8 (2018), 1053–1061, doi:10.1038/s41558-018-0305-8. On tipping point for Amazon, see T. E. Lovejoy and C. Nobre, "Amazon Tipping Point: Last Chance for Action," *Science Advances* 5 (2019), eaba2949, online, open access, https://advances.sciencemag.org /content/5/12/eaba2949; also see Anthony Faiola, Marina Lopes, and Chris Mooney, "The Price of 'Progress' in the Amazon," *Washington Post*, 28 June 2019.

45. Convention on Biological Diversity, *Global Biodiversity Outlook 5*, Summary for Policy Makers, 18 August 2020, https://www.cbd.int/gbo5. Also see Elizabeth Kolbert, *The Sixth Extinction: An Unnatural History* (New York: Bloomsbury, 2014); World Wide Fund for Nature and Zoological Society of London, *Living Planet Report 2020: Bending the Curve of Biodiversity Loss*, summary, 2020, https://livingplanet.panda.org /voices/a-life-on-our-planet; and Matthew Green, "Climate Change Could Trigger Sudden Losses of World's Wildlife: Study," *Reuters*, 8 April 2020, reporting on a study in *Nature* whose lead author Alex Pigot commented, "We found that climate change risks to biodiversity don't increase gradually. . . . It's not a slippery slope, but a series of cliff edges, hitting different areas at different times."

46. Will Steffen et al., "Trajectories of the Earth System in the Anthropocene," *Proceedings of the National Academy of Sciences of the United States of America* 115 (2018), 8252–8259, doi:10.1073/pnas.1810141115; and Juan C. Rocha et al., "Cascading Regime Shifts within and across Scales," *Science* 362 (December 2018), 1379–1383. For cascading bad human health effects, see Nick Watts et al., "The 2018 Lancet Countdown on Health and Climate Change: Shaping the Health of Nations for Centuries to Come," *Lancet* 392 (December 2018), 2479–2514, doi:10.1016/S0140–6736(18)32594–7.

47. For more extensive, powerful arguments to a similar conclusion, see Simon Caney, "Climate Change and the Future: Discounting for Time, Wealth, and Risk," *Journal of Social Philosophy* 40 (2009), 163–186. The probabilities of the melting of the Greenland Ice Sheet and the West Antarctic Ice Sheet are today, however, no longer considered "very low" (178)—or even low, as we have seen above.

48. Another example of a tipping point that, based on a quite different kind of evidence (years of sediment cores), appears already to have been passed is the Louisiana marshes on the coast of the Gulf of Mexico—see Chris Mooney, "Loss of Louisiana Marshes That Protect New Orleans Is 'Probably Inevitable,' Study Finds," *Washington Post*, 22 May 2020. The Tulane geologist who led the study of the cores is quoted as saying, "We are, if you believe this study, past the tipping point. . . . This region, it's severely wounded." Not only is the sea rising, but the land's defenses are collapsing, in part because of the oil and gas pipelines that lacerate this part of the Mississippi Delta.

49. Chris Mooney and Brady Dennis, "Ice Loss from Antarctica Has Sextupled since the 1970s, New Research Finds," *Washington Post*, 14 January 2019, reporting on Eric Rignot et al., "Four Decades of Antarctic Ice Sheet Mass Balance from

1979–2017," *Proceedings of the National Academy of Sciences of the United States of America* 116 (2019), 1095–1103, doi:10.1073/pnas.1812883116. Also see Chris Mooney, "At This Rate, Earth Risks Sea Level Rise of 20 to 30 Feet, Historical Analysis Shows," *Washington Post*, 20 September 2018, reporting on David J. Wilson et al., "Ice Loss from the East Antarctic Ice Sheet during Late Pleistocene Interglacials," *Nature* 561 (2018), 383–389, doi:10.1038/s41586-018-0501-8. And also see Julia Jacobs, "Gigantic Cavity in Antarctica Glacier Is a Product of Rapid Melting, Study Finds," *New York Times*, 1 February 2019, reporting on P. Milillo et al., "Heterogeneous Retreat and Ice Melt of Thwaites Glacier, West Antarctica," *Science Advances* 5, eaau3433 (2019), 1–8, https://advances.sciencemag.org/content/5/1/eaau3433.full.

50. Thomas Slater et al., "Review Article: Earth's Ice Imbalance," *Cryosphere* 15 (2021), 233–246, doi:10.5194/tc-15-233-2021. Also see Michael Wood et al., "Ocean Forcing Drives Glacier Retreat in Greenland," *Science Advances* 7 (2021), 1–10, doi:10.1126/sciadv.aba7282. For an accessible account of the disturbing significance of the latter research, see Chris Mooney and Andrew Freedman, "Earth Is Now Losing 1.2 Trillion Tons of Ice Each Year. And It's Going to Get Worse," *Washington Post*, 25 January 2021.

Chapter 2: The Presence of the Past

1. Toni Morrison, "The Site of Memory," in *Toni Morrison: What Moves at the Margin*, ed. Carolyn C. Denard, quoted in Emily Raboteau, "Lessons in Survival," *New York Review of Books*, 21 November 2019, 13–15, at 15.

2. Andreas Malm, *The Progress of This Storm: Nature and Society in a Warming World* (London: Verso, 2018), 96–97.

3. Václav Havel, *Letters to Olga, June 1979–September 1982*, trans. Paul Wilson (Boston and London: Faber and Faber, 1988), letter 138, 350.

4. I explain in chapter 4 why it is irrational and immoral to continue extraction-as-usual now in hope that CO_2 removal can later reverse the damage done.

5. Philippe Ciais et al., "Carbon and Other Biogeochemical Cycles," in *Climate Change 2013: The Physical Science Basis*, ed. Thomas F. Stocker et al., Contribution of Working Group I to the Fifth Assessment Report of the Intergovernmental Panel on Climate Change (Cambridge: Cambridge University Press, 2013), 465–570, at 472–473.

6. "Outcome of the 24th Session of the Conference of the Parties (COP24) to the UN Framework Convention on Climate Change (UNFCCC)," US Department of State, Office of the Spokesperson, Media Note 288121, 15 December 2018, http://www.government-world.com/press-releases-outcome-of-the-24th-session-of-the-conference-of-the-parties-cop24-to-the-un-framework-convention-on-climate-change-unfccc/.

7. I am grateful to Janina Dill for gently prodding me to be more explicit about individual and state—and their tortuous relationships.

8. For an excellent critique of some of the usual arguments, see Megan Blomfield, *Global Justice, Natural Resources, & Climate Change* (Oxford: Oxford University Press, 2019), 178–192.

9. Simon Caney has suggested that this emphasis on contribution to the problem rests on the deeper principle "that persons should take responsibility for their actions and ends"—see his "Human Rights, Responsibilities, and Climate Change," in *Global Basic Rights*, ed. Charles R. Beitz and Robert E. Goodin (Oxford: Oxford University Press, 2009), 227–247, at 241. Caney notes that this principle is an important assumption frequently invoked explicitly by John Rawls—see, for example, John Rawls, *Collected Papers*, ed. Samuel Freeman (Cambridge, MA: Harvard University Press, 1999), 261 and 284.

10. SEI (Stockholm Environment Institute) et al., *The Production Gap: The Discrepancy between Countries' Planned Fossil Fuel Production and Global Production Levels Consistent with Limiting Warming to 1.5°C or 2°C*, 2019, 8, https://productiongap.org/2019report/.

11. Health Effects Institute, *State of Global Air 2020: Special Report* (Boston: Health Effects Institute, 2020).

12. Benjamin Franta, "Early Oil Industry Knowledge of CO_2 and Global Warming," *Nature Climate Change* 8 (2018), 1024–1026, doi:10.1038/s41558-018-0349-9; Neela Banerjee, "Exxon's Oil Industry Peers Knew about Climate Dangers in the 1970s Too," *Inside Climate News*, 22 December 2015, https://insideclimatenews .org/news/22122015/exxon-mobil-oil-industry-peers-knew-about-climate-change -dangers-1970s-american-petroleum-institute-api-shell-chevron-texaco; and David Hasemyer and John H. Cushman Jr., "Exxon Sowed Doubt about Climate Science for Decades by Stressing Uncertainty," *Inside Climate News*, 22 October 2015, https:// insideclimatenews.org/news/22102015/exxon-sowed-doubt-about-climate-science -for-decades-by-stressing-uncertainty/. Also see chapter 3, notes 10 and 17; and chapter 5, notes 31 and 33.

13. Simon Caney similarly maintains that the basis for the responsibility of the inheritors is not primarily their own contribution to climate change, but their acquiring goods produced by predecessors who contributed to climate change. This is one of the continuities. In addition, many of the inheritors do also continue to be contributors to further climate change. See Caney, "Human Rights, Responsibilities, and Climate Change," at 240n53, and 244n57.

14. See Henry Shue, "Transboundary Damage in Climate Change: Criteria for Allocating Responsibility," in *Distribution of Responsibilities in International Law*, ed. André Nollkaemper and Dov Jacobs (Cambridge: Cambridge University Press, 2015), 321–340.

15. Myles R. Allen et al., "Framing and Context," in *Global Warming of 1.5 °C*, ed. Valérie Masson-Delmotte et al., Special Report (Intergovernmental Panel on Climate Change, 2018), 49–91, https://www.ipcc.ch/sr15/chapter/chapter-1/. Also see SEI et al., *Production Gap* 2019, 8.

16. I have been highlighting this distinction since my first article about climate change in 1992; see "The Unavoidability of Justice," reprinted in Henry Shue, *Climate Justice: Vulnerability and Protection* (Oxford: Oxford University Press, 2014), 41n11. In her fine book, *Global Justice, Natural Resources, & Climate Change*, Blomfield is especially alert to the significance of this distinction.

17. This is why I cannot simply invoke the powerful arguments in Daniel Butt, *Rectifying International Injustice: Principles of Compensation and Restitution between Nations* (Oxford: Oxford University Press, 2009).

18. It will also have substantial benefits, although some individuals will incur more costs than benefits. In any case, the fairness argument leaves aside relative benefit and focuses strictly on the allocation of costs.

19. See Dan Tong et al., "Committed Emissions from Existing Energy Infrastructure Jeopardize 1.5 °C Climate Target," *Nature* 572 (2019), 373–377, doi:10.1038/s41586-019-1364-3.

20. See Will Englund, "As Covid-19 Hits, Coal Companies Aim to Cut the Tax They Pay to Support Black-Lung Miners," *Washington Post*, 8 April 2020; Taylor Kuykendall "Murray Energy Executives Saw Ailing Coal Company as Own 'Piggy Bank'—Creditors," *S&P Global Market Intelligence*, 5 May 2020, https://www.spglobal.com/marketintelligence/en/news-insights/latest-news-headlines/creditors-murray-energy-execs-saw-ailing-coal-company-as-own-piggy-bank-58440831; Jeremy Hill, "Lender Says Murray Energy Violated Deal in 'Brazen' Loan Default," *Bloomberg*, 13 May 2020, https://www.bloomberg.com/news/articles/2020-05-13/lender-says-murray-energy-violated-deal-in-brazen-loan-default; and Chris Hamby, *Soul Full of Coal Dust: A Fight for Breath and Justice in Appalachia* (New York: Little, Brown, 2020).

21. Jules L. Coleman, "The Morality of Strict Tort Liability," *William and Mary Law Review* 18 (1976), 259–286, at 286. I am grateful to Catriona McKinnon for this source and for insightful comments on this issue.

22. If space permitted, one could examine a series of variants on the basic story.

23. Leslie Hook, "Climate Change: How China Moved from Leader to Laggard," *Financial Times*, 25 November 2019.

24. The objections I will explain to what I will call externalization are not grounded in inefficiency and waste, as the strictly economic objections are, but in rights violation, unfairness, and destructiveness. Below I will explain the objections separately for wrongful harms and ordinary costs. For a clear statement of the standard economic objection to externalities, see John Broome, *Climate Matters: Ethics in a Warming World* (New York: W. W. Norton, 2012), 39–40.

25. Amitav Ghosh, *The Great Derangement: Climate Change and the Unthinkable* (Chicago: University of Chicago Press, 2016), 59. He goes on to write of "unbearably intimate connections over vast gaps in time and space," having noted, "No less than they mock the discontinuities and boundaries of the nation-state do these connections defy the boundedness of 'place,' creating continuities of experience between Bengal and Louisiana, New York and Mumbai, Tibet and Alaska" (63 and 62).

26. David Archer, *The Long Thaw: How Humans Are Changing the Next 100,000 Years of Earth's Climate* (Princeton, NJ: Princeton University Press, 2009), 172–173. Quoted and perceptively interpreted in a discussion of the "genealogy of the carbon economy" by Ghosh, *Great Derangement*, 108–111.

27. It is somewhat contentious to deny that such damage is wrongful harm, and this is not a thesis to which I am attached. It can be reasonably argued that unnecessarily damaging the property of others, and especially damaging historical buildings, public

monuments, and other property belonging to society as a whole is wrong. If so, what I am labeling ordinary costs and wrongful harms would need to be distinguished instead as something like harms to nonhumans and harms to humans.

28. For brief arguments about corporate responsibility, see Henry Shue, "Responsible for What? Carbon Producer CO_2 Contributions and the Energy Transition," *Climatic Change* 144 (2017), 591–596, doi:10.1007/s10584-017-2042-9. For fuller argument, see Marco Grasso, "Towards a Broader Climate Ethics: Confronting the Oil Industry with Morally Relevant Facts," *Energy Research & Social Science* 62 (2020), 1–11, doi:10.1016/j.erss.2019.101383.

29. Recently I have briefly set the rights violations within a more general framework in Henry Shue, *Basic Rights: Subsistence, Affluence, and U.S. Foreign Policy*, 40th anniv. ed. (Princeton, NJ: Princeton University Press, 2020), chapter 8.

30. See Henry Shue, "Avoidable Necessity: Global Warming, International Fairness, and Alternative Energy" (1995), reprinted in Shue, *Climate Justice*, 89–108.

31. Mary Robinson with Caitríona Palmer, *Climate Justice: Hope, Resilience and the Fight for a Sustainable Future* (New York: Bloomsbury, 2018).

32. See Henry Shue, "Eroding Sovereignty: The Advance of Principle," in *The Morality of Nationalism*, ed. Robert McKim and Jeff McMahan (New York: Oxford University Press, 1997), 340–359, at 353–354; reprinted in Shue, *Climate Justice*, 142–161, at 156; and Henry Shue, "Share Benefits and Burdens Equitably," Working Paper (Dublin: Mary Robinson Foundation–Climate Justice, 2014).

33. I have briefly discussed this remark in "Human Rights, Climate Change, and the Trillionth Ton" (2011), reprinted in Shue, *Climate Justice*, 297–318, at 302.

34. This could change if excessive emissions were interpreted as a source of transboundary damage of the kind prohibited by customary international law. See Michael G. Faure and André Nollkaemper, "International Liability as an Instrument to Prevent and Compensate for Climate Change," *Stanford Journal of International Law*, 43A (2007), 123–179; René Lefeber, "Climate Change and State Responsibility," in *International Law in the Era of Climate Change*, ed. Rosemary Rayfuse and Shirley V. Scott (Cheltenham: Edward Elgar, 2012), 321–349; and Shue, "Transboundary Damage in Climate Change." Liability under international law for transboundary damage generally does not require fault, but only breach of a prohibition.

35. Dale Jamieson describes it this way: "The benefits from the activities that cause climate change accrue primarily to those who are members of particular political communities, while the costs are borne primarily by those who are not members of those communities. Costs are borne by those who live beyond the borders of the major emitters"; see *Reason in a Dark Time: Why the Struggle against Climate Change Failed— and What It Means for Our Future* (New York: Oxford University Press, 2014), 100.

36. I presented a briefer argument about this issue in "Global Environment and International Inequality" (1999), reprinted in Shue, *Climate Justice*, 180–194, at 182–183.

37. I discuss them in Shue, *Basic Rights*, 111–130.

38. Thucydides, *The Peloponnesian War*, trans. Richard Crawley, (New York: E. P. Dutton, 1910), book V.89. Christian Reus-Smit argues convincingly, contrary to the usual interpretation, that Thucydides meant this as the statement of a corrupt norm to

be resisted, not a natural law of political behavior to be acknowledged; see *The Moral Purpose of the State: Culture, Social Identity, and Institutional Rationality in International Relations* (Princeton, NJ: Princeton University Press, 1999), 60.

39. Leslie Hook, "China Ramps Up Coal Power in Face of Emissions Efforts," *Financial Times*, 20 November 2019. China plans to spend $30 billion to build new coal-burning power plants in other countries under the Belt and Road Initiative; see Hook, "Climate Change."

40. At the time of writing, the worst of these irresponsible utilities include Duke Energy and Southern Company; see Justin Gillis and Michael O'Boyle, "When Will Electricity Companies Finally Quit Natural Gas?," *New York Times*, 12 November 2020.

41. It also rests on a deeper philosophical mistake about the relation of domestic justice and international justice; see Henry Shue, "The Burdens of Justice," *Journal of Philosophy* 80 (1983), 600–608.

42. Shue, "Global Environment and International Inequality," 185.

43. In order to avoid secondary complications, I have in this chapter intentionally left aside the ability to pay. As Simon Caney has argued, responsibilities to other states based on causal contributions are conditional upon having the ability to pay without neglect of one's own citizens; and if action by those with responsibility based on contribution is not adequate to deal with the problem and leaves a "remainder," the ability to pay is a ground for action by others; see Caney, "Human Rights, Responsibilities, and Climate Change," 238–245. I have made some relatively minor criticisms of Caney's interpretation there of earlier arguments of mine in Henry Shue, "Historical Responsibility, Harm Prohibition, and Preservation Requirement: Core Practical Convergence on Climate Change," *Moral Philosophy and Politics* 2 (2015), 7–31 at 24–27, doi:10.1515/mop-2013-0009.

44. The term was recommended in P. J. Crutzen and E. F. Stoermer, "The 'Anthropocene,'" *IGBP Global Change Newsletter* [Stockholm] 41 (2000), 17–18, reprinted at http://www.igbp.net/news/opinion/opinion/haveweenteredtheanthropocene.5 .d8b4c3c12bf3be638a8000578.html. The suggestion was then popularized in Paul J. Crutzen, "Geology of Mankind," *Nature* 415 (2002), 23.

Chapter 3: Engagement across Distance and Engagement across Time

1. John Donne, "Meditations upon Our Human Condition, XVII," in *Devotions on Emergent Occasions, and Several Steps in my Sickness*, 2nd ed. (London: Thomas Jones, 1624), 394–395.

2. Peter U. Clark et al., "Consequences of Twenty-First-Century Policy for Multi-Millenial Climate and Sea-Level Change," *Nature Climate Change* 6 (2016), 360–369, at 360–361, doi:10.1038/nclimate2923, quoted in Andreas Malm, *The Progress of This Storm: Nature and Society in a Warming World* (London: Verso, 2018), 7, who exaggerates only slightly: "An eternity is determined now."

3. T. E. Lovejoy and C. Nobre, "Amazon Tipping Point: Last Chance for Action," *Science Advances* 5 (2019), eaba2949, https://advances.sciencemag.org/content/5/12/eaba2949.

4. I have given reasons for thinking that they do in *Basic Rights: Subsistence, Affluence, and U.S. Foreign Policy*, 40th anniv. ed. (Princeton, NJ: Princeton University Press, 2020), chapter 1.

5. Samuel Scheffler, "Individual Responsibility in a Global Age" (1995), reprinted in *Boundaries and Allegiances: Problems of Justice and Responsibility in Liberal Thought* (New York: Oxford University Press, 2001), 32–47, at 39. He also notes a third assumption that I leave aside here.

6. Geoffrey Parker, *Global Crisis: War, Climate Change & Catastrophe in the Seventeenth Century* (New Haven, CT: Yale University Press, 2014); Sam White, *A Cold Welcome: The Little Ice Age and Europe's Encounter with North America* (Cambridge, MA: Harvard University Press, 2017); and Philipp Blom, *Nature's Mutiny: How the Little Ice Age Transformed the West and Shaped the Present* (New York: W. W. Norton, 2019).

7. See Elizabeth Ashford, "The Duties Imposed by the Human Right to Basic Necessities," in *Freedom from Poverty as a Human Right: Who Owes What to the Very Poor?*, ed. Thomas Pogge (Oxford: Oxford University Press, 2007), 183–218, at 194–200; and Judith Lichtenberg, *Distant Strangers: Ethics, Psychology, and Global Poverty* (Cambridge: Cambridge University Press, 2014), 73–96.

8. For a potentially even deeper connection to self-interest, see the final paragraphs of this chapter.

9. Karen Dawisha, *Putin's Kleptocracy: Who Owns Russia?* (New York: Simon & Schuster, 2014); and Rachel Maddow, *Blowout: Corrupted Democracy, Rogue State Russia, and the Richest, Most Destructive Industry on Earth* (New York: Crown, 2019).

10. See Robert J. Brulle, "The Climate Lobby: A Sectoral Analysis of Lobbying Spending on Climate Change in the USA, 2000–2016," *Climatic Change* 149 (2018), 289–303, doi:10.1007/s10584-018-2241-z; *Big Oil's Real Agenda on Climate Change: How the Oil Majors Have Spent $1bn since Paris on Narrative Capture and Lobbying on Climate*, Influence Map, March 2019; *Update Briefing*, October 2019, https://influencemap.org/report/How-Big-Oil-Continues-to-Oppose-the-Paris-Agreement-38212275958aa21196dae3b76220bddc; and for a list of senators and members of Congress who are recipients of oil industry donations (with the amounts of the donations) and who urged special additional financial breaks for oil companies during the coronavirus pandemic, see Chris D'Angelo, "Lawmakers Pushing Big Financial Break for Oil Got $35 Million in Industry Donations," *HuffPost*, 24 March 2020, updated 3 April 2020, https://www.huffingtonpost.co.uk/entry/lawmakers-request-trump-cut-oil-gas-subsidies_n_5e7a6eefc5b6e051e8dcd458?ri18n=true; Mary Papenfuss, "Ted Cruz Got $35 Million for Billionaire Fracking Donors in Last Covid-19 Aid: Report," *HuffPost*, 29 December 2020, updated 31 December 2020, https://www.huffingtonpost.co.uk/entry/wilks-fracking-donors-ted-cruz-35-million-loan-coronavirus-relief_n_5fea86eec5b64e442105c571?ri18n=true; and Bailout Watch at https://bailoutwatch.org/.

11. For example, L. J. Sonter et al., "Renewable Energy Production Will Exacerbate Mining Threats to Biodiversity," *Nature Communications* 11 (2020), doi/10.1038/s41467-020-17928-5.

12. See, for example, Jason Gutierrez, " 'Within Seconds Everything Was Gone': Devastating Floods Submerge the Philippines," *New York Times*, 18 November 2020 (Typhoon Vamco); and Michael T. Klare, *All Hell Breaking Loose: The Pentagon's Perspective on Climate Change* (New York: Metropolitan Books, 2019), 40–48.

13. Philippe Ciais et al., "Carbon and Other Biogeochemical Cycles," in *Climate Change 2013: The Physical Science Basis*, ed. Thomas F. Stocker et al., Contribution of Working Group I to the Fifth Assessment Report of the Intergovernmental Panel on Climate Change (Cambridge: Cambridge University Press, 2013), 465–570, at 472–473.

14. David J. Frame, Adrian H. Macey, and Myles R. Allen, "Cumulative Emissions and Climate Policy," *Nature Geoscience* 7 (2014), 692–693, doi:10.1038/ngeo2254; and Luke Sussams, *Carbon Budgets Explained* (blog), Carbon Tracker Initiative, 6 February 2018, https://www.carbontracker.org/carbon-budgets-explained/.

15. See globalwarmingindex.org, accessed 5 January 2021, https://globalwarming index.org.

16. Sivan Kartha and Paul Baer, *Zero Carbon Zero Poverty the Climate Justice Way: Achieving an Equitable Phase-Out of Carbon Emissions by 2050 while Protecting Human Rights*, Report 1 2015 V1 Feb (Dublin: Mary Robinson Foundation—Climate Justice, 2015), https://www.mrfcj.org/pdf/2015-02-05-Zero-Carbon-Zero-Poverty -the-Climate-Justice-Way.pdf.

17. Benjamin Franta, "Early Oil Industry Knowledge of CO_2 and Global Warming," *Nature Climate Change* 8 (2018), 1024–1026, doi:10.1038/s41558-018-0349-9; Neela Banerjee, "Exxon's Oil Industry Peers Knew about Climate Dangers in the 1970s Too," *Inside Climate News*, 22 December 2015, https://insideclimatenews.org/news /22122015/exxon-mobil-oil-industry-peers-knew-about-climate-change-dangers -1970s-american-petroleum-institute-api-shell-chevron-texaco; and David Hasemyer and John H. Cushman Jr., "Exxon Sowed Doubt about Climate Science for Decades by Stressing Uncertainty," *Inside Climate News*, 22 October 2015. Also see chapter 5, notes 31 and 33.

18. Sivan Kartha et al., "Whose Carbon Is Burnable? Equity Considerations in the Allocation of a 'Right to Extract,'" *Climatic Change* 150 (2018; corrected 2019), 117–129, doi:10.1007/s10584-018-2209-z. Also see Greg Muttitt and Sivan Kartha, "Equity, Climate Justice and Fossil Fuel Extraction: Principles for a Managed Phase Out," *Climate Policy* 20 (2020), doi:10.1080/14693062.2020.1763900.

19. The argument that supply-side measures are essential is made in SEI (Stockholm Environment Institute) et al., *The Production Gap: The Discrepancy between Countries' Planned Fossil Fuel Production and Global Production Levels Consistent with Limiting Warming to 1.5°C or 2°C* (2019), http://productiongap.org/.

20. David Coady et al., "Global Fossil Fuel Subsidies Remain Large: An Update Based on Country-Level Estimates" (IMF Working Paper 19/89, 2019), https://www .imf.org/en/Publications/WP/Issues/2019/05/02/Global-Fossil-Fuel-Subsidies

-Remain-Large-An-Update-Based-on-Country-Level-Estimates-46509. Besides direct subsidies, the IMF includes the costs of environmental and health damage caused by fossil fuels but externalized.

21. Elizabeth Bast et al., *The Fossil Fuel Bailout: G20 Subsidies for Oil, Gas and Coal Exploration* (London: Overseas Development Institute; and Washington: OilChange International, 2014).

22. On air pollution, see Karn Vohra et al., "Global Mortality from Outdoor Fine Particle Pollution Generated by Fossil Fuel Combustion: Results from GEOS-Chem," *Environmental Research* 195 (2021), doi:10.1016/j.envres.2021.110754. On fracking pollution, see United States Environmental Protection Agency, *Hydraulic Fracturing for Oil and Gas: Impacts from the Hydraulic Fracturing Water Cycle on Drinking Water Resources in the United States (Final Report)*, EPA-600-R-16-236F (Washington, DC, 2016), https://cfpub.epa.gov/ncea/hfstudy/recordisplay.cfm?deid=332990; Andrew J. Kondash, Nancy E. Lauer, and Avner Vengosh, "The Intensification of the Water Footprint of Hydraulic Fracturing," *Science Advances* 4 (2018), eaar5982, 1–8, doi:10.1126/sciadv.aar5982; Maddow, *Blowout*, 144–151.

23. Naomi Klein, *This Changes Everything: Capitalism vs. the Climate* (London: Allen Lane, 2014), 305. Klein continues, "In an often cited statistic, an *Exxon Valdez*-worth of oil has spilled in the Delta every year for about fifty years" (305). Ken Saro-Wiwa, who was hanged by the military regime of Sani Abacha, had said, "They are going to arrest us all and execute us. All for Shell" (307). Also see Mike Corder, "Dutch Court Orders Shell Nigeria to Compensate Farmers," *AP*, 29 January 2021; and Jane Croft, Anjli Raval, and Neil Munshi, "Shell Can Be Sued in London over Alleged Nigerian Pollution," *Financial Times*, 12 February 2021.

24. Nnimmo Bassey, *To Cook a Continent: Destructive Extraction and the Climate Crisis in Africa* (Wantage, Oxon. and Nairobi: Fahamu, 2012), 164–165; Nnimmo Bassey, *We Thought It Was Oil, but It Was Blood* (2002; Ibadan: Kraft Books, Ltd., 2008); and Neil Munshi, "Graft and Mismanagement Claims Taint Nigeria Oil Clean-Up," *Financial Times*, 29 December 2019.

25. See Jane Mayer, *Dark Money: The Hidden History of the Billionaires behind the Rise of the Radical Right* (New York: Doubleday, 2016); and additional sources discussed in chapter 5.

26. Imaginative suggestions are available in Paul Hawken, ed., *Drawdown: The Most Comprehensive Plan Ever Proposed to Reverse Global Warming* (London: Penguin Random House, 2017); and Mike Berners-Lee, *There Is No Planet B: A Handbook for the Make or Break Years* (Cambridge: Cambridge University Press, 2019). Also see Reuters Staff, "IEA Says Climate Change Goals Hinge on Developing World Transition," *Reuters*, 27 January 2021.

27. That a corporation is not serious about tackling climate change is demonstrated when the financial rewards for its top executives provide incentives for greater production. See Mike Coffin and Andrew Grant, "Fanning the Flames: How Executives Continue to Be Rewarded to Produce More Oil and Gas at Odds with the Energy Transition," Carbon Tracker Initiative, 13 March 2020, https://carbontracker.org/reports/fanning-the-flames/. See also Kira Taylor, " 'Gas Is over,' EU Bank Chief

Says," EURACTIV.com, 21 January 2021, https://www.euractiv.com/section/energy-environment/news/gas-is-over-eu-bank-chief-says/.

28. See International Energy Agency (IEA), *World Energy Outlook 2020: Executive Summary*, accessed 4 April 2021, https://www.iea.org/reports/world-energy-outlook-2020#executive-summary. Also see Kingsmill Bond, *2020 Vision: Why You Should See Peak Fossil Fuels Coming*, Carbon Tracker Initiative, 10 September 2018, https://carbontracker.org/reports/2020-vision-why-you-should-see-the-fossil-fuel-peak-coming/; Tim Buckley, "IEEFA Presentation: China, Germany, South Australia, and California Are Leaders in Embracing the Global Renewable Energy Transition," Institute for Energy Economics and Financial Analysis (IEEFA), 19 February 2020), https://ieefa.org/china-germany-south-australia-and-california-leaders-in-embracing-global-renewable-energy-transition-watch/; and Tim Buckley, "Renewable Energy Could Leave USD20 Trillion of Fossil Fuel Assets Stranded within 30 Years," IEEFA, 5 March 2020, https://ieefa.org/ieefa-update-renewable-energy-could-leave-usd20-trillion-of-fossil-fuel-assets-stranded-within-30-years/.

29. Matt Gray and Sriya Sundaresan, *How to Waste over Half a Trillion Dollars: The Economic Implications of Deflationary Renewable Energy for Coal Power Investments*, Carbon Tracker Initiative, 12 March 2020, https://carbontracker.org/reports/how-to-waste-over-half-a-trillion-dollars/.

30. See Appendix on Inequality.

31. Danny Cullenward and David G. Victor, *Making Climate Policy Work* (Medford, MA: Polity Press, 2021), 9: "Relying on markets to redirect those political forces takes a hard problem and makes it even harder to solve."

32. I would count effective forms of carbon capture and storage (CCS), if they were ever developed at scale, as alternative energy: if energy is provided without allowing CO_2 to escape into the atmosphere, that counts as alternative energy. CCS so far remains undeveloped because the fossil-fuel companies have refused to invest any significant portion of their vast income into making the products they sell less polluting and harmful. If they somehow became willing to invest seriously in CCS, they might be able to rescue the now-threatened value of their reserves, which otherwise are liable to become stranded as people cease to be willing to tolerate the damage their technologically avoidable emissions do to the only planet we have. For why CCS would be necessary, see James Leaton et al., *Unburnable Carbon 2013: Wasted Capital and Stranded Assets*, Carbon Tracker Initiative and Grantham Research Institute, 2013, http://www.carbontracker.org/wp-content/uploads/2014/09/Unburnable-Carbon-2-Web-Version.pdf.

33. Tim Buckley, "Prime Minister Narendra Modi's New 'One Sun One World One Grid' Vision Positive," *IEEFA Weekly Dispatch*, 12 June 2020, https://ieefa.org/ieefa-india-prime-minister-narendra-modis-new-one-sun-one-world-one-grid-vision-positive/.

34. Buckley, "Prime Minister Narendra Modi's Vision."

35. Shue, *Basic Rights*, 13–64.

36. China has built more solar and wind power than any other country, but it is also burning more coal, by far, than any other country. On the whole, it is making

climate change much worse. Adding solar and wind on top of fossil-fuel combustion does nothing at all to slow climate change. The solar and wind must replace the fossil fuel if the carbon emissions are to be reduced. See Leslie Hook, "Climate Change: How China Moved from Leader to Laggard," *Financial Times*, 25 November 2019; and United Nations Environment Programme, *Emissions Gap Report 2020* (Nairobi: UNEP, 2020), https://www.unep.org/emissions-gap-report-2020.

37. Mengpin Ge, Johannes Friedrich, and Thomas Damassa, "6 Graphs Explain the World's Top 10 Emitters," *Climate Insights Blog*, World Resources Institute, 25 November 2014, http://www.wri.org/blog/2014/11/6-graphs-explain-world%E2%80%99s -top-10-emitters; Hannah Ritchie and Max Roser, "CO_2 and Greenhouse Gas Emissions," Our World in Data, accessed 4 April 2021, https://ourworldindata.org/co2 -and-other-greenhouse-gas-emissions. Specific country totals are at "Cumulative CO_2 Emissions," Our World in Data, accessed 4 April 2021, https://ourworldindata.org /grapher/cumulative-co-emissions?tab=chart&country=OWID_WRL; and "CAIT Climate Data Explorer," World Resources Institute, accessed 4 April 2021, https://cait .wri.org/historical/Country%20GHG%20Emissions#.

38. Henry Shue, "The Unavoidability of Justice" (1992), reprinted in Shue, *Climate Justice: Vulnerability and Protection* (Oxford: Oxford University Press, 2014), 27–46.

39. I explore what this means given the possibility of technologies of carbon dioxide removal in chapter 4.

40. Tim Carlsen et al., "Guest Post: How Declining Ice in Clouds Makes High 'Climate Sensitivity' Plausible," *Carbon Brief*, 30 October 2020, https://www.carbonbrief .org/guest-post-how-declining-ice-in-clouds-makes-high-climate-sensitivity -plausible; Timothy M. Lenton et al., "Climate Tipping Points—Too Risky to Bet Against," *Nature* 575 (2019; corrected 2020), 592–595, at 595, doi:10.1038/d41586-019-03595-0; and Will Steffen et al., "Trajectories of the Earth System in the Anthropocene," *Proceedings of the National Academy of Sciences of the United States of America* 115 (2018), 8252–8259, doi:10.1073/pnas.1810141115.

41. R. T. Pierrehumbert, "Cumulative Carbon and Just Allocation of the Global Carbon Commons," *Chicago Journal of International Law* 13 (2013), 527–548; Matthew Collins et al., "Long-Term Climate Change: Projections, Commitments and Irreversibility," in *Climate Change 2013: The Physical Science Basis*, ed. Thomas F. Stocker et al., Contribution of Working Group I to the Fifth Assessment Report of the Intergovernmental Panel on Climate Change (Cambridge: Cambridge University Press, 2013), 1029–1136; and Frame, Macey, and Allen, "Cumulative Emissions and Climate Policy."

42. China's Ministry of Environment and Ecology is making some effort to discourage Chinese banks from providing the financing for polluting BRI projects; see Christian Shepherd, "Belt and Road Pollution Blacklist Discourages Fossil Fuel Investments," *Financial Times*, 1 December 2020.

43. Similar international schemes are being proposed in Australia and China; see Odysseus Patrick, "The $16 Billion Plan to Beam Australia's Outback Sun onto Asia's Power Grids," *Washington Post*, 10 August 2020; and James Kynge and Lucy Hornby, "China Eyes Role as World's Power Supplier," *Financial Times*, 7 June 2018.

44. Elizabeth Cripps, "Population and Environment: The Impossible, the Impermissible, and the Imperative," in *The Oxford Handbook of Environmental Ethics*, ed. Stephen M. Gardiner and Allen Thompson (New York: Oxford University Press, 2017), 380–390.

45. I originally called the exploitation of vulnerabilities that one has contributed to creating "compound injustice" in Shue, "Unavoidability of Justice," 36–41.

46. Bryan Denton and Somini Sengupta, "India's Ominous Future: Too Little Water, or Far Too Much," *New York Times*, 25 November 2019.

47. Henry Shue, "Subsistence Emissions and Luxury Emissions" (1993), reprinted in Shue, *Climate Justice*, 47–67, at 52–53.

48. Matthew 27:24 (Revised Standard Version).

49. Tim Buckley and Charles Worringham, "India's Electricity Sector Transformation Has Made Progress in 2019/20," IEEFA India, 28 November 2019, https://ieefa .org/ieefa-india-indias-electricity-sector-transformation-has-made-progress-in-2019 -20/. "Investors continue to pivot away from Indian thermal power generation due to renewable energy being the lowest cost source of new power generation, and the ongoing shortages in cost-effective fossil fuels continuing."

50. Shue, *Basic Rights*, 119–127.

51. Darrel Moellendorf, *The Moral Challenge of Dangerous Climate Change: Values, Poverty, and Policy* (Cambridge: Cambridge University Press, 2014), 22.

52. Practical Action, *Poor People's Energy Outlook 2013: Energy for Community Services* (Rugby, UK: Practical Action Publishing Ltd., 2013), x. On Africa's need for investment to avoid carbon lock in, see Kim Harrisberg, "Africa's Leap to Clean Energy Seen Needing Policy Change, Investment," *Reuters*, 11 January 2021, and the underlying study reported on.

53. Pope Francis, *Laudato Si': Encyclical Letter of the Holy Father Francis on Care for Our Common Home* (2015), para. 162, 120, http://w2.vatican.va/content/francesco /en/encyclicals/documents/papa-francesco_20150524_enciclica-laudato-si.html. An earlier paragraph had observed, "Intergenerational solidarity is not optional, but rather a basic question of justice, since the world we have received also belongs to those who will follow us" (para. 159, 118).

54. Henry Shue, "Deadly Delays, Saving Opportunities: Creating a More Dangerous World?" (2010), reprinted in Shue, *Climate Justice*, 263–286; and Lauren Hartzell-Nichols, *A Climate of Risk: Precautionary Principles, Catastrophes, and Climate Change* (New York: Routledge, 2017), 44–82.

55. See discussion in chapter 1, note 26.

56. Lenton et al., "Climate Tipping Points," 595.

57. Thomas Sumner, "No Stopping the Collapse of West Antarctic Ice Sheet," *Science* 344 (2014), 683; Ian Joughin, Benjamin E. Smith, and Brooke Medley, "Marine Ice Sheet Collapse Potentially Under Way for the Thwaites Glacier Basin, West Antarctica," *Science* 344 (2014), 735–738, doi:10.1126/science.1249055; and E. Rignot et al., "Widespread, Rapid Grounding Line Retreat of Pine Island, Thwaites, Smith, and Kohler Glaciers, West Antarctica, from 1992 to 2011," *Geophysical Research Letters* 41 (2014), 3502–3509, doi:10.1002/2014GL060140.

58. Lovejoy and Nobre, "Amazon Tipping Point." Also see Henry Fountain, "'Going in the Wrong Direction': More Tropical Forest Loss in 2019," *New York Times*, 2 June 2020; and Jake Spring, "Exclusive: Brazil Amazon Fires Likely Worst in 10 Years, August Data Incomplete, Government Researcher Says," *Reuters*, 2 September 2020.

59. For a graphic, interactive, highly accessible, and extensive presentation of the underlying dynamics of AMOC, see Moises Velasquez-Manoff and Jeremy White, "In the Atlantic Ocean, Subtle Shifts Hint at Dramatic Dangers," *New York Times*, 2 March 2021, https://www.nytimes.com/interactive/2021/03/02/climate/atlantic -ocean-climate-change.html. For details, see L. Caesar et al., "Current Atlantic Meridional Overturning Circulation Weakest in Last Millenium," *Nature Geoscience* 14 (2021), 118–120, doi:10.1038/s41561-021-00699-z; and L. Caesar et al., "Observed Fingerprint of a Weakening Atlantic Ocean Overturning Circulation," *Nature* 556 (2018), 191–196, doi:10.1038/s41586-018-0006-5. Also see Chris Mooney, "The Oceans' Circulation Hasn't Been This Sluggish in 1,000 Years. That's Bad News," *Washington Post*, 11 April 2018, https://www.washingtonpost.com/news/energy-environment/wp/2018 /04/11/the-oceans-circulation-hasnt-been-this-sluggish-in-1000-years-thats-bad-news /?utm_term=.3be3b0282343.

60. Henry Fountain, "Alaska's Permafrost Is Thawing," *New York Times*, 23 August 2017; and Boris K. Biskaborn et al., "Permafrost Is Warming at a Global Scale," *Nature Communications* 10 (2019), doi:10.1038/s41467-018-08240-4.

61. Alexey Portnov et al., "Offshore Permafrost Decay and Massive Seabed Methane Escape in Water Depths >20m at the South Kara Sea Shelf," *Geophysical Research Letters* 40 (2013), 3962–3967, doi:10.1002/grl.50735.

62. A research team led by scientists from Ohio State University believes that it has evidence that the melting of the Greenland ice is already irreversible: "We show that widespread retreat between 2000 and 2005 resulted in a step-increase in discharge and a switch to a new dynamic state of sustained mass loss that would persist even under a decline in surface melt"; see Michaela D. King et al., "Dynamic Ice Loss from the Greenland Ice Sheet Driven by Sustained Glacier Retreat," *Communications Earth & Environment* 1 (2020), abstract, doi:10.1038/s43247-020-0001-2.

63. Scheffler, *Boundaries and Allegiances*, 39.

64. Luke 10:29–37 (Revised Standard Version).

65. Toby Ord notices this but does not seem fully to appreciate its significance; see *The Precipice: Existential Risk and the Future of Humanity* (New York: Bloomsbury Publishing, 2020), 386n34.

66. Catriona McKinnon, *Climate Change and Future Justice: Precaution, Compensation, and Triage* (Abingdon: Routledge, 2012), 129.

67. Anja Karnein, "Putting Fairness in Its Place: Why There Is a Duty to Take Up the Slack," *Journal of Philosophy* 111 (2014), 593–607.

68. Thanks, of course, to Bernard Williams.

69. I am grateful to Joshua Goldstein for raising this issue.

70. Jonathan Schell, *The Fate of the Earth* (New York: Alfred A. Knopf, 1982), part 2, "The Second Death," 117.

71. Samuel Scheffler, *Death & the Afterlife*, ed. Niko Kolodny (New York and Oxford: Oxford University Press, 2013), 30 and 26.

72. Hannah Arendt, "Europe and the Atom Bomb" (1954), reprinted in *Essays in Understanding, 1930–1954: Formation, Exile, and Totalitarianism*, ed. Jerome Kohn (New York: Schocken Books, 1994), 421–422.

73. The remaining paragraphs in this chapter draw on Henry Shue, "Uncertainty as the Reason for Action: Last Opportunity and Future Climate Disaster," in "Global Justice and Climate," special issue, *Global Justice: Theory Practice Rhetoric* 8 (2016), 86–103, https://www.theglobaljusticenetwork.org/global/index.php/gjn/article/view/89/65.

74. I am grateful to Anja Karnein for discussion of this issue.

75. Provided of course that the actions we choose to take are not misguided or counterproductive. But we have overwhelming evidence that carbon emissions need to be phased out before the cumulative carbon budget for an endurable temperature is exceeded, so there is no danger that exiting fossil fuels rapidly is a mistake.

Chapter 4: Are There Second Chances on Climate Change?

1. Professor Katharine Hayhoe, Twitter, https://twitter.com/KHayhoe/status/1242817345069998080, quoted in Meehan Crist, "What the Coronavirus Means for Climate Change," *New York Times*, 27 March 2020.

2. Professor Cristian Proistosescu, Twitter, https://twitter.com/cristiproist/status/1304211038980759553, quoted in John Branch and Brad Plumer, "Climate Disruption Is Now Locked In. The Next Moves Will Be Crucial," *New York Times*, 22 September 2020.

3. Ada Limón, "Salvage," Greenpeace, February 2020, https://www.greenpeace.org/usa/stories/ada-limon-on-our-climate-in-crisis, quoted in Michaela Coplen et al., "Theme Section Introduction," *St Antony's International Review* 15 (2020), 3–7, at 3.

4. Henry Shue, "Climate Dreaming: Negative Emissions, Risk Transfer, and Irreversibility," *Journal of Human Rights and Environment* 8 (2017), 203–216, doi:10.4337/jhre.2017.02.02.

5. Dominic Lenzi, "The Ethics of Negative Emissions," *Global Sustainability* 1 (2018), 1–8, doi:10.1017/sus.2018.5; and "On the Permissibility (or Otherwise) of Negative Emissions," *Ethics, Policy & Environment* (2021), published online 5 February 2021, doi:10.1080/21550085.2021.1885249.

6. Jan C. Minx et al., "Negative Emissions—Part 1: Research Landscape and Synthesis," *Environmental Research Letters* 13 (2018), 063001, doi:10.1088/1748-9326/aabf9h.

7. Nathaniel Rich argues that the best opportunity to prevent the worst arose in the previous decade but was similarly wasted through corporate deception and political irresponsibility; see *Losing Earth: The Decade We Could Have Stopped Climate Change* (New York: Picador, 2019).

8. David Wallace-Wells, *The Uninhabitable Earth: A Story of the Future* (New York: Tim Duggan Books, 2019), 4.

9. Hannah Ritchie and Max Roser, "CO$_2$ and Greenhouse Gas Emissions," Our World in Data, accessed 4 April 2021, https://ourworldindata.org/co2-and-other -greenhouse-gas-emissions. These specific totals were retrieved at "Per Capita CO$_2$ Emissions, 2019," Our World in Data, accessed 4 April 2021, https://ourworldindata .org/grapher/cumulative-co-emissions?tab=chart&country=OWID. Cumulative emissions from the United States rose from 256.95 billion tonnes in 1992 to 399.38 billion tonnes in 2017, while cumulative emissions from China rose from 47.00 billion to 200.14. Thus, cumulative Chinese emissions in 2017 were lower than cumulative US emissions in 1992 and only slightly more than half US emissions in 2017, for a vastly larger population. These figures are exclusively for emissions from fossil fuels and cement and omit emissions from agriculture, forestry, and other land-use changes, which add significantly to the complete total.

10. Elizabeth Kolbert, "What Will Another Decade of Climate Crisis Bring?," *New Yorker*, 13 January 2020, https://www.newyorker.com/magazine/2020/01/13/what -will-another-decade-of-climate-crisis-bring.

11. Timothy Mitchell, "Afterword to the Paperback Edition," in *Carbon Democracy: Political Power in the Age of Oil* (New York: Verso, 2013), 260.

12. Dieter Helm, *Net Zero: How We Stop Causing Climate Change* (London: William Collins, 2020), 1.

13. Chris Mooney, "Scientists Descended into Greenland's Perilous Ice Caverns— and Came Back with a Worrying Message," *Washington Post*, 23 December 2020.

14. SEI (Stockholm Environment Institute) et al., *The Production Gap: The Discrepancy between Countries' Planned Fossil Fuel Production and Global Production Levels Consistent with Limiting Warming to 1.5°C or 2°C* (2019), 8, http://productiongap.org/.

15. G. P. Peters et al., "Carbon Dioxide Emissions Continue to Grow amidst Slowly Emerging Climate Policies," *Nature Climate Change* 10 (2020), 3–6, doi:10.1038/ s41558-019-0659-6. See especially figure 2.

16. See chapter 5, notes 9 and 10. This social phenomenon is somewhat similar to the kinds of recalcitrant "rebound" phenomena clearly described in Mike Berners-Lee and Duncan Clark, *The Burning Question: We Can't Burn Half the World's Oil, Coal and Gas. So How Do We Quit?* (London: Profile Books, 2013), 47–63.

17. Christiana Figueres et al., "Three Years to Safeguard Our Climate," *Nature* 546 (2017), 593–595.

18. Christian Holz, Sivan Kartha, and Tom Athanasiou, "Fairly Sharing 1.5: National Fair Shares of a 1.5°C-Compliant Global Mitigation Effort," *International Environmental Agreements: Politics, Law and Economics* 18 (2018), 117–134, at 131, doi:10.1007/ s10784-017-9371-z.

19. Pete Smith et al., "Biophysical and Economic Limits to Negative CO$_2$ Emissions," *Nature Climate Change* 6 (2016), 42–50, at 43, doi:10.1038/nclimate2870. The notation "CO$_2$eq" indicates the incorporation of the effects of GHGs other than CO$_2$ as the amount of CO$_2$ that would produce that same effect—the equivalent amount of CO$_2$, or CO$_2$eq.

20. Myles Allen et al., "Summary for Policymakers," in *Global Warming of 1.5 °C*, ed. Valérie Masson-Delmotte et al., Special Report (Intergovernmental Panel on Climate Change, 2018), C.3, 17, https://www.ipcc.ch/sr15/chapter/spm/.

21. I am grateful to Catriona McKinnon for rescuing me from even clumsier labels.

22. See the discussion in chapter 5.

23. Alex Lenferna, "Divest-Invest: A Moral Case for Fossil Fuel Divestment," in *Climate Justice: Integrating Economics and Philosophy*, ed. Ravi Kanbur and Henry Shue (Oxford: Oxford University Press, 2019), 139–156.

24. Kejun Jiang et al., "Mitigation Pathways Compatible with 1.5°C in the Context of Sustainable Development," in Masson-Delmotte et al., *Global Warming of 1.5 °C*, 2.3.4.1, 122. In the caption for figure 2.9, the first and second purposes are labeled "compensatory CO_2" and "net negative CO_2"; the first label seems confusing to me.

25. The important argument that failure to carry out ambitious mitigation now increases a nation's responsibility to conduct compensating remedial CDR later is made, for the first time of which I am aware, in Claire L. Fyson et al., "Fair-Share Carbon Dioxide Removal Increases Major Emitter Responsibility," *Nature Climate Change* 10 (2020), 836–841, doi:10.1038/s41558-020-0857-2.

26. Kevin Anderson and Glen Peters, "The Trouble with Negative Emissions," *Science* 354 (2016), 182–183, doi:10.1126/science.aah4567; Oliver Geden and Andreas Löschel, "Define Limits for Temperature Overshoot Targets," *Nature Geoscience* 10 (2017), 881–882, doi:10.1038/s41561-017-0026-z; and Oliver Geden, "Politically Informed Advice for Climate Action," *Nature Geoscience* 11 (2018), 380–383, doi:10.1038/s41561-018-0143-3.

27. Edward A. Parson and Holly J. Buck, "Large-Scale Carbon Dioxide Removal: The Problem of Phasedown," *Global Environmental Politics* 20 (2020), 70–92, at 71.

28. For further complications, see Parson and Buck, "Large-Scale Carbon Dioxide Removal."

29. Minx et al., "Negative Emissions—Part 1," abstract and 17–18.

30. Jiang et al., "Mitigation Pathways Compatible," 2.3.4, 121. See also Brad Plumer and Christopher Flavelle, "Businesses Aim to Pull Greenhouse Gases from the Air. It's a Gamble," *New York Times*, 18 January 2021.

31. A delightful and highly accessible account of all this is in Tyler Volk, *CO₂ Rising: The World's Greatest Environmental Challenge* (Cambridge, MA: MIT Press, 2008). Every earthling—and especially beer-drinkers—should read this little paperback.

32. See the searchable database at "Explore the Data," Forests & Finance, https://forestsandfinance.org/data/.

33. Jiang et al., "Mitigation Pathways Compatible," 2.3.4; 2.3.4.1; 2.3.4.2; and Heleen de Conick et al., "Strengthening and Implementing the Global Response," in *Global Warming of 1.5 °C*, ed. Valérie Masson-Delmotte et al., Special Report (Intergovernmental Panel on Climate Change, 2018), 4.3.7.

34. Melissa Chan and Heriberto Araújo, "China Wants Food. Brazil Pays the Price," *The Atlantic*, 15 February 2020.

35. Smith et al., "Biophysical and Economic Limits to Negative CO_2 Emissions," 47 and 46.

36. Rob Bellamy and Oliver Geden, "Govern CO_2 Removal from the Ground Up," *Nature Geoscience* 12 (2019), 874–879, at 874.

37. I did not take this into account in "Climate Dreaming," as observed in Minx et al., "Negative Emissions—Part I," 19.

38. Minx et al., "Negative Emissions—Part I," abstract and 13; Jiang et al., "Mitigation Pathways Compatible," 2.3.4.1, 122.

39. Bellamy and Geden, "Govern CO_2 Removal from the Ground Up."

40. I will simply mention a few concerns. Because SRM could affect the people of every nation on the planet, it ought to be employed only after an international consensus has been reached on a system of global governance for its implementation and maintenance, not imposed on everyone else by the wealthiest and most technologically advanced nations pursuing their own national interests. SRM does nothing to change the mechanisms that are causing climate change but instead blocks (only?) the step of temperature rise late in the process by—if it works as hoped—blocking the right amount of incoming solar radiation; in this respect it deals with only the harmful symptoms, but not the underlying disruption to climate dynamics. It does nothing to reduce the acidification of the oceans or the rampant lung and heart disease that currently result from the air pollution from the combustion of fossil fuel. And it would nevertheless strongly tempt fossil-fuel interests to continue to resist a transition away from the current energy regime and to continue business-as-usual. Most important at present, while matters are urgent, they are not desperate. Sophisticated modeling shows multiple pathways by which net zero emissions can be reached in time if we can retire the politicians who are currently blocking for the fossil-fuel industry. For three examples for the United States, see National Academies of Sciences, Engineering, and Medicine, *Accelerating Decarbonization of the U.S. Energy System* (Washington, DC: National Academies Press, 2021), doi:10.17226/25932; James H. Williams et al., "Carbon-Neutral Pathways for the United States," *AGU Advances* 2 (2021), doi:10.1029/2020AV000284; and Eric Larson et al., *Net-Zero America: Potential Pathways, Infrastructure, and Impacts*, interim report, 15 December 2020 (v2) (Princeton, NJ: Andlinger Center, Princeton University), https://netzeroamerica.princeton .edu/the-report.

41. A third thesis that I believe is a natural extension of (2) has already been introduced in chapter 1 and supported independently there and in chapter 3: (3) most pointedly, mitigation that is so lacking in ambition that it bequeaths risks that remain unlimited, when the risks could without inordinate sacrifice have been at least limited, is intolerably objectionable (and snatches defeat from the jaws of a grand historic opportunity to bring climate change under control).

42. Héiène Hermansson and Sven Ove Hansson, "A Three-Party Model Tool for Ethical Risk Analysis," *Risk Management* 9 (2007), 129–144, at 131.

43. Jonathan Wolff, "Five Types of Risky Situation," *Law, Innovation and Technology* 2 (2010), 151–163, at 159.

44. John Rawls, *A Theory of Justice*, rev. ed. (Cambridge, MA: Belknap Press of Harvard University Press, 1999), 134; [1st ed. (1971), 154–155]; and Stephen M. Gardiner, "A Core Precautionary Principle," *Journal of Political Philosophy* 14 (2006), 33–60.

45. Rawls, *Theory of Justice*, rev. ed., 134 [1st ed., 154].

46. David A. Weisbach, "The Problems with Climate Ethics," in Stephen M. Gardiner and David A. Weisbach, *Debating Climate Ethics* (New York: Oxford University Press, 2016), 135–244, at 172.

47. The same applies among the three generations that together constitute what I am calling "the present generation"; the inaction of the grandfathers has inflicted an unreasonable gamble on the grandchildren alive today. This was one basis of the "children's lawsuit," *Juliana vs. the United States*; see Wallace-Wells, *Uninhabitable Earth*, 167.

48. Personal communication, April 2019.

49. Deirdre Cooper, "Low Rates Provide a Historic Opportunity to Tackle Climate Change," *Financial Times*, 26 December 2019.

50. Ian Joughin, Benjamin E. Smith, and Brooke Medley, "Marine Ice Sheet Collapse Potentially Under Way for the Thwaites Glacier Basin, West Antarctica," *Science* 34 (2014) 4, 735–738, doi:10.1126/science.1249055; and E. Rignot et al., "Widespread, Rapid Grounding Line Retreat of Pine Island, Thwaites, Smith, and Kohler Glaciers, West Antarctica, from 1992 to 2011," *Geophysical Research Letters* 41 (2014), 3502–3509, doi:10.1002/2014GL060140.

51. For a fuller but still accessible account of the mechanisms underlying the melting, see British Antarctic Survey, "Deep Channels Link Ocean to Antarctic Glacier," press release, 9 September 2020, https://www.bas.ac.uk/media-post/deep-channels -link-ocean-to-antarctic-glacier/. Links are provided to the peer-reviewed technical publications summarized.

52. Yes, the glacier is sliding uphill as it approaches the sea, because the mountains of ice in WAIS weigh against it from behind.

53. Stephen Rich Rintoul et al., "Ocean Heat Drives Rapid Basal Melt of the Totten Ice Shelf," *Science Advances* 2 (2016), 1–5, https://advances.sciencemag.org/content /2/12/e1601610.

54. More recently, warming ocean water has been found in addition to be undercutting East Antarctica's Denman Glacier by penetrating the deepest continental gorge on earth, which happens to lie beneath Denman; see V. Brancato et al., "Grounding Line Retreat of Denman Glacier, East Antarctica, Measured with COSMO-SkyMed Radar Interferometry Data," *Geophysical Research Letters* 47 (2020), 1–10, doi:10.1029/2019GL086291. For an account accessible to laypersons, see Chris Mooney, "Scientists Just Discovered a Massive New Vulnerability in the Antarctic Ice Sheet," *Washington Post*, 23 March 2020. Denman consists of water equivalent to a global sea-level rise of 1.5 m and appears to be in rapid retreat.

55. Annette Baier, "Poisoning the Wells," in *Values at Risk*, ed. Douglas MacLean, Maryland Studies in Public Philosophy (Totowa, NJ: Rowman & Allanheld, 1986), 49.

56. See chapter 1 and, for example, Gregory S. Cooper, Simon Willcock, and John A. Dearing, "Regime Shifts Occur Disproportionately Faster in Larger Ecosystems," *Nature Communications* 11 (2020), 1–10, doi:10.1038/s41467-020-15029-x. They conclude, "The collapse of large vulnerable ecosystems, such as the Amazon rainforest and Caribbean coral reefs, may take only a few decades once triggered" (abstract) and "humanity now needs to prepare for changes in ecosystems that are faster than

we previously envisaged through our traditional linear view of the world, including across Earth's largest and most iconic ecosystems, and the social-ecological systems that they support" (7). Also see Jeff Berardelli, "Climate Tipping Points May Have Been Reached Already, Experts Say," CBS News, 26 April 2021, https://www.cbsnews.com /news/climate-change-tipping-points-amazon-rainforest-antarctic-ice-gulf-stream/.

57. William Blake, "Milton," Book the First, 22, ll. 24–25, in *Blake: Complete Writings*, ed. Geoffrey Keynes (Oxford: Oxford University Press, 1969), 505. Blake meant a different level of permanence, however.

58. In tribute to Billy Kwan (played by Linda Hunt), *The Year of Living Dangerously*.

Chapter 5: Taking Control of Our Legacy

1. Leah Cardamore Stokes, *Short Circuiting Policy: Interest Groups and the Battle over Clean Energy and Climate Policy in the American States* (Oxford: Oxford University Press, 2020), 4, 5, and 34.

2. Timothy Mitchell, "Afterward to the Paperback Edition," in *Carbon Democracy: Political Power in the Age of Oil* (New York: Verso, 2013), 267.

3. Erik Olin Wright, *Envisioning Real Utopias* (New York and London: Verso, 2010), 294.

4. Henry Shue, "Responsible for What? Carbon Producer CO_2 Contributions and the Energy Transition," *Climatic Change* 144 (2017), 591–596, doi:10.1007/s10584-017-2042-9.

5. For documentation, see chapter 3, notes 10 and 17, and chapter 5, notes 31 and 33.

6. See, for example, Mitchell, *Carbon Democracy*; Leif Wenar, *Blood Oil: Tyrants, Violence, and the Rules That Run the World* (Oxford: Oxford University Press, 2016); Matto Mildenberger, *Carbon Captured: How Business and Labor Control Climate Politics* (Cambridge, MA: MIT Press, 2020); and Stokes, *Short Circuiting Policy*.

7. For amazingly detailed practical guidance, see Tara Shine, *How to Save Your Planet One Object at a Time* (New York: Simon & Schuster, 2020). Also see Mike Berners-Lee, *There Is No Planet B: A Handbook for the Make or Break Years* (Cambridge: Cambridge University Press, 2019); and Paul Hawken, ed., *Drawdown: The Most Comprehensive Plan Ever Proposed to Reverse Global Warming* (London: Penguin Random House, 2017).

8. At a more general level, I may be presupposing something like what Elizabeth Cripps calls "promotional duties"—"duties to attempt to bring about the necessary collective action"—for which she provides sophisticated and systematic argument; see her *Climate Change and the Moral Agent: Individual Duties in an Interdependent World* (Oxford: Oxford University Press, 2013), 140–150, at 140. Also see the taxonomy of responses in Simon Caney, "Climate Change and Non-Ideal Theory: Six Ways of Responding to Non-Compliance," in *Climate Justice in a Non-Ideal World*, ed. Clare Heyward and Dominic Roser (Oxford: Oxford University Press, 2016), 21–42.

9. G. P. Peters et al., "Carbon Dioxide Emissions Continue to Grow amidst Slowly Emerging Climate Policies," *Nature Climate Change* 10 (2020), 3–6, doi:10.1038/s41558-019-0659-6; see especially figure 2. Also see R. B. Jackson, Corinne Le Quéré,

and R. M. Andrew, "Global Energy Growth Is Outpacing Decarbonization," *Environmental Research Letters* 13 (2018), doi:10.1088/1748–9326/aaf303.

10. David Wallace-Wells, *The Uninhabitable Earth: A Story of the Future* (New York: Tim Duggan Books, 2019), 177.

11. For a report on GHG that includes those from the manufacture of plastics and other chemical processes, see Environmental Integrity Project, *Greenhouse Gases from Oil, Gas, and Petrochemical Production*, 8 January 2020, https://www.environmentalintegrity.org/reports/greenhouse-gases-from-oil-gas-and-petrochemical-production/.

12. James Bruggers, "Market Headwinds Buffet Appalachia's Future as a Center for Petrochemicals," *Inside Climate News*, 21 March 2020, https://insideclimatenews.org/news/21032020/appalachia-future-center-petrochemicals-coronavirus-plastic-ethane/.

13. Laura Sullivan, "Plastic Wars: Industry Spent Millions Selling Recycling—To Sell More Plastic," NPR, 31 March 2020, https://www.npr.org/2020/03/31/822597631/plastic-wars-three-takeaways-from-the-fight-over-the-future-of-plastics.

14. Hiroko Tabuchi, Michael Corkery, and Carlos Mureithi, "Big Oil Is in Trouble. Its Plan: Flood Africa with Plastic," *New York Times*, 30 August 2020.

15. Nicholas Kusnetz, "In Louisiana, Stepping onto Oil and Gas Industry Land May Soon Get You 3 Years or More in Prison," *Inside Climate News*, 10 June 2020, https://insideclimatenews.org/news/10062020/louisiana-petrochemical-plant-environmental-justice/.

16. Robert Schuwerk and Greg Rogers, *It's Closing Time: The Huge Bill to Abandon Oilfields Comes Early*, 18 June 2020 (London: Carbon Tracker Initiative, 2020), 11 and 36, https://carbontracker.org/reports/its-closing-time/.

17. Hiroko Tabuchi, "Fracking Firms Fail, Rewarding Executives and Raising Climate Fears," *New York Times*, 12 July 2020 (updated 17 July). Contrast Kristina Marusic, "Fractured," Part 1: "Harmful Chemicals and Unknowns Haunt Pennsylvanians Surrounded by Fracking," *Environmental Health News*, 1 March 2021, https://www.ehn.org/fractured-harmful-chemicals-fracking-2650428324.html.

18. Schuwerk and Rogers, *Closing Time*, 10. ARO for a single especially deep fracked well can cost $1 million.

19. Mya Frazier, "Gas Companies Are Abandoning Their Wells, Leaving Them to Leak Methane Forever," *Bloomberg Green*, 17 September 2020, https://www.bloomberg.com/news/features/2020-09-17/abandoned-gas-wells-are-left-to-spew-methane-for-eternity. Employment-creating ways to tackle the results of this corporate ruthlessness are being formulated: Jason Bordoff, Daniel Raimi, and Neelesh Nerurkar, *Green Stimulus for Oil and Gas Workers: Considering a Major Federal Effort to Plug Orphaned and Abandoned Wells*, Resources for the Future and Center on Global Energy Policy, Columbia University, 20 July 2020, https://www.energypolicy.columbia.edu/research/report/green-stimulus-oil-and-gas-workers-considering-major-federal-effort-plug-orphaned-and-abandoned.

20. The Carbon Tracker Initiative in London coined the terms "stranded assets" and "stranded liabilities," I believe. Carbon Tracker continues to produce perceptive

and penetrating reports on the fossil-fuel industry, as does the Institute for Energy Economics and Financial Analysis in Cleveland. Each has valuable studies available for free online.

21. Kingsmill Bond, Ed Vaughan, and Harry Benham, *Decline and Fall: The Size and Vulnerability of the Fossil Fuel System*, Carbon Tracker Initiative, 4 June 2020, 24, https://carbontracker.org/reports/decline-and-fall/. Also see Fiona Harvey, "State-Owned Fossil Fuel Firms' Plan to Invest $1.9tn Could Destroy Climate Hopes," *Guardian*, 9 February 2021: "National oil companies (NOCs) produce about two-thirds of the world's oil and gas and own about 90% of reserves."

22. Bond, Vaughan, and Benham, *Decline and Fall*, 8. Shareholders are becoming increasingly restive—see Attracta Mooney, "Shareholder Climate Rebellions Surge despite Coronavirus Crisis," *Financial Times*, 30 May 2020.

23. Clifford Kraus, " 'I'm Just Living a Nightmare': Oil Industry Braces for Devastation," *New York Times*, 21 April 2020.

24. On ExxonMobil's deep entanglements with Putin's Rosneft, see Rachel Maddow, *Blowout: Corrupted Democracy, Rogue State Russia, and the Richest, Most Destructive Industry on Earth* (New York: Crown, 2019), 269–289. More generally, see Karen Dawisha, *Putin's Kleptocracy: Who Owns Russia?* (New York: Simon & Schuster, 2014); and Steve Coll, *Private Empire: ExxonMobil and American Power* (New York: Penguin, 2012).

25. Charles Koch's decision not to go public has enabled him to maintain an extreme degree of secrecy about his operations; for example, see the account given by Christopher Leonard, *Kochland: The Secret History of Koch Industries and Corporate Power in America* (New York: Simon & Schuster, 2019), 133–134, about Koch's ability to get away with the destruction of internal documents sought by a federal prosecutor in the case of oil allegedly stolen from Native Americans' wells. One of Koch's often-invoked guiding principles is said to be "the whale that comes above sea level gets harpooned" (458). Also see Mayer, *Dark Money*.

26. For BlackRock's extensive support for coal-burning utilities, see Europe Beyond Coal, *Fool's Gold: The Financial Institutions Risking Our Renewable Energy Future with Coal*, 15 July 2020, 11ff., https://beyond-coal.eu/2020/07/15/fools-gold-2020/. Cf. Attracta Mooney, "BlackRock Punishes 53 Companies over Climate Inaction," *Financial Times*, 14 July 2020.

27. Political contributions in the United States are tracked in the *OpenSecrets Newsletter* published by the Center for Responsive Politics, https://www.opensecrets.org/.

28. David Coady et al., "Global Fossil Fuel Subsidies Remain Large: An Update Based on Country-Level Estimates" (IMF Working Paper 19/89, 2019), https://www.imf.org/en/Publications/WP/Issues/2019/05/02/Global-Fossil-Fuel-Subsidies-Remain-Large-An-Update-Based-on-Country-Level-Estimates-46509. Besides direct subsidies, the IMF includes the costs of environmental and health damage caused by fossil fuels but externalized. Also see Elizabeth Bast et al., *Empty Promises: G20 Subsidies to Oil, Gas and Coal Production* (Washington, DC: Oil Change International, 2015), http://priceofoil.org/2015/11/11/empty-promises-g20-subsidies-to-oil-gas-and-coal-production/; Alex Doukas, *Talk Is Cheap: How G20 Governments Are Financing*

Climate Disaster (Washington, DC: Oil Change International, 2017), http://priceofoil
.org/content/uploads/2017/07/talk_is_cheap_G20_report_July2017.pdf; and Global
Gas and Oil Network, *Oil, Gas and the Climate: An Analysis of Oil and Gas Industry Plans
for Expansion and Compatibility with Global Emission Limits*, 2019, www.ggon.org. And
note Stokes, *Short Circuiting Policy*, 15: "While fossil fuel subsidies are permanent,
these renewable energy tax credits expire."

    29. Nadja Popovich, Livia Albeck-Ripka, and Kendra Pierre-Louis, "The Trump
Administration Is Reversing 100 Environmental Rules. Here's the Full List," *New
York Times*, 20 January 2021; and Juliet Eilperin, Brady Dennis, and John Muyskens,
"Trump Rolled Back More Than 125 Environmental Safeguards. Here's How," *Washington Post*, 30 October 2020.

    30. Coady et al., "Global Fossil Fuel Subsidies Remain Large," abstract.

    31. The Russian corruption is definitively documented in Dawisha, *Putin's Kleptocracy*. Also see Catherine Belton, *Putin's People: How the KGB Took Back Russia and Then
Took On the West* (London: William Collins, 2020). The Russian federal government is
especially dependent on revenue from fossil fuels because of the Putin regime's failure
to diversify the economy: "The oil and gas sector is estimated to contribute between
10% and 20% of Russia's GDP and almost half of federal government revenues"—
SEI (Stockholm Environment Institute) et al., *The Production Gap: The Discrepancy
between Countries' Planned Fossil Fuel Production and Global Production Levels Consistent with Limiting Warming to 1.5°C or 2°C* (2019), 32, https://productiongap.org
/2019report/. I do not know a study of the United States fully comparable to Dawisha,
but revealing studies include Robert J. Brulle, "Institutionalizing Delay: Foundation
Funding and the Creation of U.S. Climate Change Counter-Movement Organizations,"
Climatic Change 122 (2014), 681–694, doi:10.1007/s10584-013-1018-7; Robert J. Brulle,
"The Climate Lobby: A Sectoral Analysis of Lobbying Spending on Climate Change
in the USA, 2000–2016," *Climatic Change* 149 (2018), 289–303, doi:10.1007/s10584-
018-2241-z; Robert J. Brulle, "Networks of Opposition: A Structural Analysis of U.S.
Climate Change Countermovement Coalitions, 1989–2015," *Sociological Inquiry* 20
(2019), 1–22; Alexander Hertel-Fernandez, *State Capture: How Conservative Activists, Big Businesses, and Wealthy Donors Reshaped the American States—and the Nation*
(Oxford: Oxford University Press, 2019); Leonard, *Kochland*; Mayer, *Dark Money*; and
Theda Skocpol and Alexander Hertel-Fernandez, "The Koch Network and Republican
Party Extremism," *Perspectives on Politics* 14 (2016), 681–699. One comprehensive
philosophical analysis of the nature of the corruption is Wenar, *Blood Oil*. Both Coll,
Private Empire, and Maddow, *Blowout*, have revealing studies of ExxonMobil's role
in Equatorial Guinea, one of the most corrupt (and cruel) governments on earth,
which is also discussed in Wenar. A careful case study of the importance of accelerated
depreciation of new capital investment, as provided by the US intangible drilling cost
subsidy, is Peter Erickson et al., "Why Fossil Fuel Producer Subsidies Matter," *Nature*
578, E1–E4, doi:10.1038/s41586-019-1920-x. The last concludes, "The economic, political and symbolic effects of subsidies reinforce each other."

    32. For detailed accounting, see Dan L. Wagner et al., *Bailed Out & Propped Up:
U.S. Fossil Fuel Pandemic Bailouts Climb toward $15 Billion*, Bailout Watch, Public

Citizen and Friends of the Earth, November 2020, https://report.bailoutwatch.org/. Also see, for example, Amy Harder, "U.S. Spends the Most Stimulus but Smallest Share on Green Energy," *Axios*, 8 September 2020.

33. Center for International Environmental Law, *Smoke and Fumes: The Legal and Evidentiary Basis for Holding Big Oil Accountable for the Climate Crisis* (Washington, DC: CIEL, 2017), 7–25, https://www.ciel.org/reports/smoke-and-fumes/. A classic study of corporate antiscience strategy is Naomi Oreskes and Erik M. Conway, *Merchants of Doubt: How a Handful of Scientists Obscured the Truth on Issues from Tobacco Smoke to Global Warming* (New York: Bloomsbury Press, 2010), 169–215. An account by a leading scientist attacked by the industry is Michael E. Mann, *The Hockey Stick and the Climate Wars: Dispatches from the Front Lines* (New York: Columbia University Press, 2012). Coll's magisterial study, *Private Empire*, summarizes with elegant understatement: "with its ideological allies, ExxonMobil funded the promotion of public confusion about climate science by means that future employees and executives of the corporation are likely to look back on with regret" (620). Also see the sources in chapter 3, notes 10 and 17.

34. David Anderson, Matt Kasper, and David Pomerantz, *Utilities Knew: Documenting Electric Utilities Early Knowledge and Ongoing Deception on Climate Change from 1968–2017* (San Francisco: Energy and Policy Institute, 2017), https://www.energyandpolicy.org/utilities-knew-about-climate-change/. For utilities' expenditures on lobbying, see Brulle, "Climate Lobby."

35. Hiroko Tabuchi, "How One Firm Drove Influence Campaigns Nationwide for Big Oil," *New York Times*, 11 November 2020; Zach Boren, Alexander C. Kaufman, and Lawrence Carter, "Revealed: BP and Shell Back Anti-Climate Lobby Groups despite Pledges," *HuffPost*, 28 September 2020; and Justin Mikulka, "Major Fossil Fuel PR Group Is behind Europe Pro-Hydrogen Push," *DeSmog: Clearing the PR Pollution That Clouds Climate Science*, 9 December 2020, https://www.desmogblog.com/2020/12/09/fti-consulting-fossil-fuel-pr-group-behind-europe-hydrogen-lobby.

36. Wallace-Wells, *Uninhabitable Earth*, 220.

37. David Sheppard, "Pandemic Crisis Offers Glimpse into Oil Industry's Future," *Financial Times*, 3 May 2020.

38. Stokes, *Short Circuiting Policy*, 2.

39. Tom Sanzillo, "Here's Why the Texas Railroad Commission Should Regulate Flaring in the Oil Fields," *IEEFA Weekly Dispatch*, 12 June 2020, https://ieefa.org/ieefa-update-heres-why-the-texas-railroad-commission-should-regulate-flaring-in-the-oil-fields/.

40. Jonathan Watts, Jillian Ambrose, and Adam Vaughan, "Oil Companies Scrambling to Raise Output in Final 'Fossil Fuel Harvest,'" *Guardian*, 11 October 2019, Special Investigation: The Polluters, 12. Also see Stokes, *Short Circuiting Policy*, 240.

41. Decisions about whose assets are left in the ground ought, of course, to be made democratically, not left to oil executives or state bureaucrats. For examination of whose assets ought to be the ones stranded, see Greg Muttitt and Sivan Kartha, "Equity, Climate Justice and Fossil Fuel Extraction: Principles for a Managed Phase Out," *Climate Policy* 20 (2020), doi:10.1080/14693062.2020.1763900. Fair phase-out is a relatively unexplored topic.

42. Michael Lazarus and Cleo Verkuijl, *The Production Gap 2020 Special Report: The Discrepancy between Countries' Planned Fossil Fuel Production and Global Production Levels Consistent with Limiting Warming to 1.5°C or 2°C* (SEI, IISD, ODI, E3G, and UNEP, 2020), 3 and figure ES.1, http://productiongap.org/2020report. Chapter 4 discusses issues in a just transition away from fossil-fuel production, especially support for badly affected workers and communities. The website includes supporting appendices. An authoritative *Emissions Gap* report has been published annually for a number of years; see, for example, UNEP, UNEP DTU Partnership, *Emissions Gap 2020* (Nairobi: UN Environment Programme, 2020), https://www.unep.org/emissions-gap-report -2020. *Production Gap* reports will now appear annually as well.

43. Lazarus and Verkuijl, *Production Gap 2020 Special Report*, 22. Also see Erickson et al., "Why Fossil Fuel Producer Subsidies Matter."

44. Lazarus and Verkuijl, *Production Gap 2020 Special Report*, 25. On the US government specifically, see Wagner et al., *Bailed Out & Propped Up*; and Harder, "U.S. Spends the Most Stimulus but Smallest Share on Green Energy."

45. A somewhat jargon-ridden but still instructive study of BP's advertising at an earlier period is Julie Doyle, "Where Has All the Oil Gone? BP Branding and the Discursive Elimination of Climate Change Risk," in *Culture, Environment and Ecopolitics*, ed. Nick Heffernan and David Wragg (Newcastle upon Tyne: Cambridge Scholars Press, 2011), 200–225. Doyle shows how BP's introduction of the notion of a carbon footprint shifts the burden for action from the seller to the consumer, just as the petrochemical manufacturers of plastics have made the problem seem to be lack of sufficient recycling by consumers rather than far too much production and marketing by the plastic makers.

46. International Energy Agency, *The Oil and Gas Industry in Energy Transitions*, 20 January 2020, 47, https://webstore.iea.org/the-oil-and-gas-industry-in-energy -transitions.

47. Also see Andrew Grant and Mike Coffin, *Breaking the Habit—Why None of the Large Oil Companies Are "Paris-Aligned," and What They Need to Do to Get There*, Carbon Tracker Initiative, 13 September 2019, https://www.carbontracker.org/reports /breaking-the-habit/.

48. Sierra Club, "Citigroup Is Latest Major US Bank to Rule Out Funding for Arctic Refuge Drilling," 20 April 2020, https://www.sierraclub.org/press-releases/2020 /04/breaking-citigroup-latest-major-us-bank-rule-out-funding-for-arctic-refuge. But also see Nicholas Kusnetz, "Banks' Vows to Restrict Loans for Arctic Oil and Gas Development May Be Largely Symbolic," *Inside Climate News*, 5 May 2020, https:// insideclimatenews.org/news/05052020/oil-gas-banks-arctic-drilling-coronavirus/.

49. Alison Kirsch et al., *Banking on Climate Change: Fossil Fuel Finance Report 2020* (Rainforest Action Network et al., 2020), 3, https://www.ran.org/bankingon climatechange2020/; interactive data sites included.

50. Kirsch et al., *Banking on Climate Change*, 8. Also see Joe Romm, "The Stunning Hypocrisy of JP Morgan and CEO Jamie Dimon," *ThinkProgress*, 22 March 2019, https://archive.thinkprogress.org/stunning-hypocrisy-of-jp-morgan-jamie-dimon-on -climate-change-a3fd2ecfbbbd/.

51. Kirsch et al., *Banking on Climate Change*, 11.

52. Kirsch et al., *Banking on Climate Change*, 4.

53. Kirsch et al., *Banking on Climate Change*, 14.

54. One imaginative coalition is Stop the Money Pipeline, https://stopthe moneypipeline.com/. BankTrack provides a free weekly newsletter on climate-relevant bank activity: https://www.banktrack.org/. Bank loans to pulp and paper corporations that are fueling deforestation are now tracked at Forests & Finance, forests andfinance.org.

55. Lenferna, "Divest-Invest."

56. For a variety of views on how much and which aspects of capitalism need to change, see Leo Barasi, *The Climate Majority: Apathy and Action in an Age of National-ism* (Oxford: New Internationalist Publications, Ltd., 2017); Naomi Klein, *This Changes Everything: Capitalism vs. the Climate* (London: Allen Lane, 2014); Geoff Mann and Joel Wainwright, *Climate Leviathan: A Political Theory of Our Planetary Future* (London: Verso, 2018); and Timothy Mitchell, *Carbon Democracy: Political Power in the Age of Oil* (New York: Verso, 2013). For climate action by US Evangelical Christians, see Katharine K. Wilkinson, *Between God & Green: How Evangelicals Are Cultivating a Middle Ground on Climate Change* (New York: Oxford University Press, 2012).

57. Mildenberger, *Carbon Captured*, 236, 237, and 240; see especially 235–251. Thanks to Simon Caney for alerting me to this analysis.

58. Maddow, *Blowout*, 351–360. Also see Wenar, *Blood Oil*, 316–319.

59. Bond, Vaughan, and Benham, *Decline and Fall*, 6. Compare the report of con-clusions by influential Rystad Energy: Myles McCormick, "Coronavirus Will Hasten 'Peak Oil' by Three Years, Says Research Firm," *Financial Times*, 18 June 2020. Rystad predicts peak oil in 2027/2028.

60. Bond, Vaughan, and Benham, *Decline and Fall*, 26 and 30. Also see Interna-tional Energy Agency (IEA), *World Energy Outlook 2020: Executive Summary*, Octo-ber 2020, 18–19, https://www.iea.org/reports/world-energy-outlook-2020#.

61. Tim Buckley, "Prime Minister Narendra Modi's New 'One Sun One World One Grid' Vision Positive," *IEEFA Weekly Dispatch*, 12 June 2020, https://ieefa.org /ieefa-india-prime-minister-narendra-modis-new-one-sun-one-world-one-grid-vision -positive/.

62. Kathy Hipple, Clark Williams-Derry, and Tom Sanzillo, *Living beyond Their Means: Cash Flows of Five Oil Majors Can't Cover Dividends, Buybacks* (Cleveland: Institute for Energy Economics and Financial Analysis, 2020), figure 2, https://ieefa .org/ieefa-report-oil-majors-live-beyond-their-means-%e2%80%92-cant-pay-for -dividends-buybacks/.

63. Dino Grandoni, "Big Oil Just Isn't as Big as It Once Was," *Washington Post*, 4 September 2020.

64. Eric Platt, "ExxonMobil Booted from the Dow after Close to a Century," *Financial Times*, 25 August 2020. Chevron is the only "supermajor" left in the Dow Jones Average.

65. Kingsmill Bond et al., "The Future's Not in Plastics: Why Plastics Demand Won't Rescue the Oil Sector," Carbon Tracker Initiative, 4 September 2020, 3,

https://carbontracker.org/reports/the-futures-not-in-plastics/. Also see Pew Charitable Trusts and SYSTEMIQ, *Breaking the Plastic Wave: A Comprehensive Assessment of Pathways towards Stopping Ocean Plastic Pollution*, 2020, https://www.systemiq.earth/breakingtheplasticwave/.

66. Bond et al., "The Future's Not in Plastics," 10.

67. Bond et al., "The Future's Not in Plastics," 16.

68. Matthew Vincent, "Threat from Climate Change to Financial Stability Bigger Than Covid-19," *Financial Times*, 7 June 2020. Also see Finance Watch, *Report—Breaking the Climate-Finance Doom Loop*, 8 June 2020, https://www.finance-watch.org/publication/breaking-the-climate-finance-doom-loop/.

69. Coral Davenport and Jeanna Smialek, "Federal Report Warns of Financial Havoc from Climate Change," *New York Times*, 8 September 2020. Also see Nick Cunningham, "'Uninsurable' and 'Unhedgeable': Central Banks Warn of Financial Crisis from Climate Change," *DeSmog: Clearing the PR Pollution That Clouds Climate Science*, 31 January 2020; and Marilyn Waite, "Carbon Accounting Should Be a Basic Requirement for Banks," *Financial Times*, 30 August 2020.

70. Wenar, *Blood Oil*, 285–288.

71. Klein, *This Changes Everything*, 389.

72. Mitchell, *Carbon Democracy*, 23–27.

73. Jonathan Schell, *The Unconquerable World: Power, Nonviolence, and the Will of the People* (New York: Penguin, 2005), 144. Schell wrote, "Violence is a method by which the ruthless few can subdue the passive many. Nonviolence is a means by which the active many can overcome the ruthless few." A beautiful example is the successful opposition by the Augusta County Alliance (now part of the Alliance for the Shenandoah Valley), Chesapeake Climate Action Network, Cowpasture River Preservation Association, and numerous other grassroots organizations, in concert with the Southern Environmental Law Center in Charlottesville, against the destructive and unnecessary Atlantic Coast Pipeline for fracked gas that Dominion Energy and Duke Energy worked for years to force on them.

74. See, for example, Dieter Helm, *Burn Out: The Endgame for Fossil Fuels* (New Haven, CT: Yale University Press, 2017); and Kingsmill Bond, *2020 Vision: Why You Should See Peak Fossil Fuels Coming*, Carbon Tracker Initiative, 10 September 2018, https://carbontracker.org/reports/2020-vision-why-you-should-see-the-fossil-fuel-peak-coming/. Also see Global Commission on the Geopolitics of Energy Transformation, *A New World: The Geopolitics of the Energy Transformation* (International Renewable Energy Agency, 2019), https://www.irena.org/publications/2019/Jan/A-New-World-The-Geopolitics-of-the-Energy-Transformation.

75. For a concrete proposal, see Stuart Jenkins et al., "Sustainable Financing of Permanent CO_2 Disposal through a Carbon Takeback Obligation," https://arxiv.org/abs/2007.08430.

76. Some obvious minimum steps for foreign policy are outlined by John Podesta and Todd Stern, "A Foreign Policy for the Climate: How American Leadership Can Avert Catastrophe," *Foreign Affairs*, May/June 2020. Two critical ones are to work closely with India in avoiding China's coal-intensive form of development and instead

building at last major parts of the One Sun One World One Grid electricity system and to work with allies to prevent the Bolsonaro Administration in Brazil from pushing the Amazon still further into an "ecological death spiral."

77. Walt Kelly, *Pogo: We Have Met the Enemy and He Is Us* (New York: Simon and Schuster, 1972), 11.

Appendix on Inequality

1. See Leif Wenar, *Blood Oil: Tyrants, Violence, and the Rules That Run the World* (Oxford: Oxford University Press, 2016).

2. An accessible overview is available in an Oxfam Media Briefing: Tim Gore with Mira Alestig and Anna Ratcliff, "Confronting Carbon Inequality: Putting Climate at the Heart of the COVID-19 Recovery," Oxfam, 21 September 2020, https://www.oxfam .org/en/research/confronting-carbon-inequality. The full report, with details about methodology and the dataset, is S. Kartha et al., *The Carbon Inequality Era: An Assessment of the Global Distribution of Consumption Emissions among Individuals from 1990 to 2015 and beyond*, Joint Research Report (Oxfam and SEI, 2020), https://www.sei .org/publications/the-carbon-inequality-era.

3. Gore with Alestig and Ratcliff, "Confronting Carbon Inequality," 2, and figures 1, 2, and 6.

4. Gore with Alestig and Ratcliff, "Confronting Carbon Inequality," 4, emphasis in original.

5. Various ways of construing this distinction are discussed in Henry Shue, "Subsistence Protection and Mitigation Ambition: Necessities, Economic and Climatic," *British Journal of Politics and International Relations*, 21 (2019), 251–262, https://doi .org/10.1177/1369148118819071.

6. Gore with Alestig and Ratcliff, "Confronting Carbon Inequality," 9.

7. Yannick Oswald, Anne Owen, and Julia K. Steinberger, "Large Inequality in International and Intranational Energy Footprints between Income Groups and across Consumption Categories," *Nature Energy* 5 (2020), 231–239, doi:10.1038/ s41560-020-0579-8.

8. Philippe Benoit, "A Luxury Carbon Tax to Address Climate Change and Inequality: Not All Carbon Is Created Equal," *Ethics & International Affairs* blog, March 2020, https://www.ethicsandinternationalaffairs.org/2020/a-luxury-carbon -tax-to-address-climate-change-and-inequality-not-all-carbon-is-created-equal/.

accelerated depreciation, 168n31
action, immediate: bank financing and, 129–130; CDR versus mitigation and, 97–103; duty and, 47; future generations and, 12, 75–76; human equality and, 79–80; international inequalities and responsibilities and, 76; rapid scaling up and, 98–103; science and, 6–7, 8–9, 21–22; self-interested reasons and, 7–8; socio-political situation and, 16–17; and that means now, 135–136; unlimited danger and, 22–24, 76, 86–87, 163n41. *See also* agency; duties: promotional; Energy Revolution; historical context and historic opportunity; inaction; pivotal (present) generation; scaling up, rapid; second chances
adaptation: date-of-last-opportunity and, 6; future generations and, 76, 85; historical context and, 5; international inequality and responsibility and, 124; sea-level rise and, 49; sovereign externalization and, 50, 72; temporal externalization and, 114
Africa, 74, 121. *See also* Niger Delta
agency: generations and, 79–80; local versus global, 60, 69, 79, 150n25; responsibility and, 60–61. *See also* action, immediate; individual agency, responsibility and rights; duties: promotional; pure fairness argument; responsibility, generational; social movements
agriculture and agribusiness: BECCS and, 101–102; Brazil and, 99; causal

responsibility and, 33; CO_2 and, 37; fossil-fuel combustion and, 6; industrialization of, 5; Little Ice Age and, 20–21; trees, land, and water and, 99–100. *See also* food
air carbon capture and storage, 100
air pollution, 33
Alaska, 79
Alliance for the Shenandoah Valley, 172n73
alternative/noncarbon/renewable energy: affordability of, 66–75, 110; avoidable necessity and, 47; avoidance of danger versus savings and, 14; CCS and, 156n32; CO_2 emissions rise and, 120; coronavirus pandemic and, 126–127; cost-sharing and, 40–41; fossil fuel energy and, 7–8; India and, 158n49; infrastructure and transmission of, 124; investments in, 66–67, 110; new global electricity and, 132; prices of, 67, 70–71, 72–73, 129, 132; 2050 and, 63. *See also* Energy Revolution; solar energy; wind energy
Amazon rainforest, 78, 81, 99, 164n56. *See also* Brazil
AMOC (Atlantic Meridional Overturning Circulation), 78–79, 159n59
Antarctic ice sheets, 24, 49, 147nn44 and 49, 164nn51 and 54; East Antarctic Ice Sheet (EAIS), 25, 112–113; West Antarctic Ice Sheet (WAIS), 25, 59, 76–79, 81, 82, 85, 111–113, 147n52
Anthropocene period, 18, 54–55, 145n29
apocalypses, 21

A NOTE ON THE TYPE

This book has been composed in Adobe Text and Gotham. Adobe Text, designed by Robert Slimbach for Adobe, bridges the gap between fifteenth- and sixteenth-century calligraphic and eighteenth-century Modern styles. Gotham, inspired by New York street signs, was designed by Tobias Frere-Jones for Hoefler & Co.